Salomat Gelchinova
Ulugbek Khudanov

Intensification of cement grinding process

Salomat Gelchinova
Ulugbek Khudanov

Intensification of cement grinding process

Monograph

ScienciaScripts

Imprint

Any brand names and product names mentioned in this book are subject to trademark, brand or patent protection and are trademarks or registered trademarks of their respective holders. The use of brand names, product names, common names, trade names, product descriptions etc. even without a particular marking in this work is in no way to be construed to mean that such names may be regarded as unrestricted in respect of trademark and brand protection legislation and could thus be used by anyone.

Cover image: www.ingimage.com

This book is a translation from the original published under ISBN 978-620-7-46542-2.

Publisher:
Sciencia Scripts
is a trademark of
Dodo Books Indian Ocean Ltd. and OmniScriptum S.R.L publishing group

120 High Road, East Finchley, London, N2 9ED, United Kingdom
Str. Armeneasca 28/1, office 1, Chisinau MD-2012, Republic of Moldova, Europe
Printed at: see last page
ISBN: 978-620-7-76818-9

Contents

The monograph is recommended for publication by the protocol of 28 December #4 2023 by the Council of Jizzak State Pedagogical University.

INTRODUCTION

Relevance of the problem. One of the main directions in the development of the national economy of the country is to ensure a worldwide growth in the efficiency of social production, improving the quality of products, strengthening the regime of economy in the national economy.

The construction materials industry of the Republic of Uzbekistan today is actively developing due to the dynamic growth of construction. It represents one of the basic components of the economy of Uzbekistan and is of crucial importance for its stable development in the long term, consumers of construction materials are practically all branches of industry, transport, agriculture and others.

In 2016, investments in fixed capital in the Republic of Uzbekistan as a whole increased by 24.8% compared to 2015. Due to the growth of investment in the economy and development of construction, cement consumption and production in the country increased significantly. A total of 8,462,000 tonnes of cement were produced in 2016, with cement production capacity increasing by 880,000 tonnes compared to 2015.

Uzbekistan currently has 12 cement plants with a total annual capacity of more than 9.0 million tonnes, including the major ones - JSC Kizilkumcement (Navoi, production capacity - 3,500,000 tonnes), JSC Ahangarancement (Ahangaran, 1,740,000 tonnes), JSC Bekabadcement (Bekabad, 1,250,000 tonnes), JSC Kuvasaycement (Kuvasay, 1,080,000 tonnes), JSC Jizabadcement (Kuvasay, 1,080,000 tonnes), and JSC Jizabad Cement (Bekabad, 1,250,000 tonnes). Bekabadcement JSC (Bekabad, 1,250,000 tonnes), Kuvasaycement JSC (Kuvasay, 1,080,000 tonnes), Jizzak Cement Plant (Jizzak, 1,000,000 tonnes of grey cement or 150,000 tonnes of white cement), - and small cement plants with a total capacity of 640,000 tonnes per year.

The capacity utilisation of the cement plants of the republic is close to 100 % y. It should be noted that at the newly introduced technological lines of JSC "Bekabadcement" and Jizzak Cement Plant the efforts of specialists of these enterprises have achieved the design indicators of cement production during a short period of operation. similar results were achieved by the teams of small enterprises of JV LLC "Fergana Cement", LLC "Turon Eco Cement Group" and LLC "Farhadshifer".

Along with increasing the volume of cement output, much attention is paid to expanding the range of products and their quality. All major cement enterprises have introduced a quality management system and obtained international certificates ISO 9001:2000 and ISO 9001:2015. Small enterprises have also started work in this direction.

In this monograph a method of obtaining surfactants - intensifiers of cement clinker grinding on the basis of cotton oil production waste - gossypol resin (cotton tar), modifying them with triethanolamine and studying the influence of new surfactants on the grinding process and cement properties has been developed.

The method of intensification of grinding of Portland cement clinker and components of raw cement sludge with application of new surfactants is offered. The optimum variant of surfactants application in the processes of Portland cement production is established.

History of the birth of the industry in the world

In December 1824, Joseph Aspdin, a stonemason from Leeds, England, was awarded patent 7 for an "Improved Method of Making Artificial Stone" which he created in his own kitchen. The inventor heated a mixture of well-detailed limestone and clay in a kitchen oven, then crushed the lump of the mixture into powder and obtained hydraulic cement, which hardened when water was added. The idea of the possibility of making from an artificial mixture of limestone and clay "a cement which would be equal to the best market Portland cement stone in hardness and durability" belongs to J. Smeaton. J. Aspdin was the first to give the hydraulic binder the name "Portland cement" (because he used stones from a quarry on the island of Portland in its manufacture) and is the inventor of the name. In 1825.

Aspdin established a factory production of his "Portland cement" in

Yorkshire, near Leeds. The major builders of the time appreciated the new cement and favoured it over all other cements, so its production developed in England. The binder obtained by Aspdin was not Portland cement in the modern sense of the word (Aspdin did not bring the mixture to sintering, which is the main condition for obtaining Portland cement clinker), but was a type of Roman cement obtained at a slightly higher firing temperature ($900\text{-}1000^{0}$ C), but the name "Portland cement" has been retained to this day. The hydraulic binder described by E. G. Cheliyev, was closer in properties to modern Portland cement, and in quality was superior to Aspdin's Portland cement, but Aspdin founded a factory production of cement (Fig. 4), which was developed by his son William [1].

Figure 1. J. Aspdin's furnace at the Robins & Aspdin factory he built, 1847-1848. [1].

All Portland cement on the continent became known as English cement until 1878-1880, when its production was mastered in Germany. From that time cement began to be used almost everywhere. The technology of cement production became widespread, and in each state industrialists built their own cement plant, and even more than one. It became much easier to buy cement, it was no longer necessary to transport it by sea. Aspdin's sons

William and James first produced "Aspdin" cement at their established plants, and in 1848 Robins, Aspdin and Company began producing real Portland cement. In France, Lafarge began producing Portland cement for the first time from 1868 at Tey, commune of Ardèche. Although the first cement plant in France was built in 1846 in Boulogne-sur-Mer, it produced the then popular Roman cement [1].

The first Portland cement factory in Germany was established in Zülchow8 near Stettin in 1855. The cement, produced from local clay and chalk from the island of Wolin, was as good as English cement, and production soon increased to 30,000 barrels a year. This success prompted the establishment of similar factories in Bonn, Lebbing, Oppeln, Lüneburg, Amenenburg, and Finkelwald. By 1877, there were already 30 Portland cement plants in Germany. Cement plants (cement factories) appeared everywhere where large deposits of limestone and clay were found. For example, in Germany in 1859, while exploring lignite deposits in the suburb of Hemmoor, they came across thick layers of chalk and clay. In 1862, Jürgen-Hinrich Hagenach built a brickworks and a lime factory on the site, and in 1866 a small cement factory, Hemmoor, was established. In 1882 the joint-stock company "Portland cement factory Hemmoor" with 1000 workers produced 170 000 barrels of cement per year. The first cement plant in the United States, Coplay Cement Company, was founded in 1866 by David O. Saylor in the Lehigh Valley of Pennsylvania. David O. Saylor is considered the "father" of the American cement industry. This particular cement plant, now a museum, was owned by Saylor, and of course the museum was created in his honour.

Lehigh County is a natural source of raw materials for cement production. In 1871 Saylor received a patent for the production of Portland cement, which was much stronger and of higher quality than the so-called natural cements produced in the United States up to that time. Saylor's cement was used to build bridges, roads, skyscrapers, aqueducts, office and residential buildings. In 1900, the Lehigh Valley produced 72% of all Portland cement in the United States. At first blast furnaces were built, but they were extremely inefficient and were quickly shut down. As early as 1893, the kilns that we can still see today, i.e. the Schoefer vertical kilns, were built (Fig. 5). Eleven vertical kilns were used for cement production, which were 90 ft10 (27.4 m) high. The Lichaj Valley kilns, a Danish modification of the German vertical kilns, were built of local red brick.

Figure 2. Coplay Company cement kilns [1].

Compared to the dome furnaces, these furnaces (continuous furnaces) produced a higher quality product. However, they did not last long either. In 1904, the vertical furnaces were closed down, as more efficient and productive horizontal rotary kilns became available [1].

Figure 3. The first rotary kiln [1].

The inventors of the rotary kiln are Elliot and Roussel, who obtained a patent for a rotary mechanical soda kiln in 1853 (Fig. 3). The invention of rotary kilns was important not only for the progress of soda production but also in other industries, including the cement industry. The mechanical rotary kiln was driven by a steam engine and had a capacity of 196.5 poods of raw charge.

The rotary kiln was first used in the cement industry by Frederick Ransom (1818 -1893; patent in England 1885, in the USA 1886). In Europe, "lying" rotary kilns instead of "standing" shaft kilns were introduced by the most famous builders F. L. Smidt in Copenhagen and A. G. E. Polisius in Dessau. The rotary kiln was a refractory-lined metal cylinder resting on roller supports. To move the fired material in the furnace it was installed at an angle of 4 -5 to the horizon. The furnace was driven by an electric motor. The raw material mixture, loaded from the cold end, moved during the rotation of the furnace towards the combustion products of the fuel. One of the first kilns was 11 metres long and had a diameter of 1.5 metres, and in 1900 there were kilns with a diameter of 2 metres and a length of 35 metres. Their daily capacity was 30 tonnes (one of the longest furnaces 232 x 7.6 m with a capacity of up to 3000 tonnes per day was installed by FLSmidth in 1964 at Dundee Clacksville, USA). The crusher (mill) for grinding clinker was invented in 1892 by Davidsen, a Danish engineer. Inside it was lined with quartz tiles, and sea pebbles were used as grinding bodies. Cement was widely advertised in the world press, and the public stopped seeing the new material as something strange and frightening. And after the

British Army and Navy had literally adopted it, the generals of all more or less developed countries became interested in cement.

The Germans started independent development of cement production, as the British were exporting the finished product, but not the secrets of its production technology.

Origin of the industry in Russia. The industrial production of cement in Russia has a century-long history. If we take as a starting point the first official mention of it, it dates back to the XVII century. In a letter to Prince M.P. Gagarin, then commandant of Moscow, Peter the Great instructed to send several barrels of lime. It is noteworthy that later the word "lime" was crossed out and corrected to "cement". It is quite possible that in this case we were talking about one of the varieties of cement, namely Roman cement, produced in those distant times, because the first cement plant in Russia was built and put into operation in 1856, and it produced Portland cement [1].

In 1822, Prof. A. R. de Charleville published scientific research on marl rocks for the production of hydraulic lime and cements at the St. Petersburg Railway Institute. The author pointed out that when firing such rocks or mixtures of limestone and clay, chemical interactions between the constituent parts occur. In 1839, merchant and manufacturer I.V. Yunker founded a factory in St. Petersburg for the production of "Parker's" or "English" cement. The factory or "English Cement Factory" was located near the old Moskovskaya Zastava and worked on "cement stone" delivered from England. In 1839 and 1840 1 and 5 thousand barrels of Yunker's cement of 10 poods each were produced respectively. Junker's Roman cement was widely used not only in St. Petersburg, but also in Moscow and was highly appreciated by builders. Later similar cement was produced from local raw materials at the factories of P.E. Roche near St. Petersburg (1848) and Filatiev near Moscow (1849), which was of great importance for construction. Roche's factory worked for 57 years and produced excellent Roman cement (at first Volkhov limestone was used for cement production, and then limestone was taken from Tosnya quarries), which was used to replace English Portland cement, which cost three times more expensive, in various underwater and onshore constructions with great economic effect. The first Russian description of a binding substance obtained by burning marl with subsequent grinding was given in 1807 by Academician V.M. Severgin. The obtained product was better in quality than Roman cement, but it was not Portland cement. The inventor of the present cement is considered to be E. G. Cheliyev. Egor Gerasimovich Cheliev started working in Saratov. In 1801 he moved to Moscow and a few years later was appointed head of the Moscow military-working brigade for the planning and restoration of Moscow buildings after the fire of 1812. It was then that he began experimenting with different materials to find a binding compound for brick and stone [1].

Fig. 4. a - a barrel with cement from the Cement Museum of Novorossiysk. The barrel bears the inscription: "Akts. Obshch. Novorossiysk Portland Cement Plant 'Chain'"; b - token of the "Partnership of Riga Cement Plant and Creamery of K.H. Schmidt". The token bears the inscription "Alexei Mikhailovich Powalishin31. 1868. In memory of the millionth barrel of cement production. 1884"; c - loading cement on the ship [1].

By the end of the 19th century, the Russian cement industry fully met the needs of construction in cement. The quality of Russian Portland cement was high. Each factory had a laboratory and controlled the products. In addition, the factories provided cement samples for testing to well-equipped departmental laboratories. Thus, the chemical composition of cement from P.E. Roche's factory was tested in 1862 by the laboratory of the mining department in St. Petersburg. Cement from the same plant was tested by A. R. Shulyachenko in 1869. On the initiative of cement plant owners, cement industry congresses were organised in 1885. The work of the congresses was firstly led by A. R. Shulyachenko and then by A. R. Shulyachenko. R. Shulyachenko, and then N.A. Belelyubsky.

Although the first rotary kilns, 25 metres long and 1.8 metres in diameter, manufactured at the Makeyevka Machine-Building Plant, appeared in Russia in 1909, clinker was produced mainly in batch and continuous shaft kilns. The labour conditions of the workers were extremely difficult. By 1927-1928 cement production in Russia had risen to the pre-war level. During the first five-year period 15 new cement plants were built. In 1940, 5.7 million tonnes of cement were produced. Two plants were built in the east of the country.

Chapter 1

Main part,
1.1 Cement production technology

Cement is a hydraulic binder obtained by fine grinding of Portland cement clinker with gypsum and additives, which, when mixed with water, forms a workable dough capable of hardening in water and in air. Cement production involves two steps: the first is the production of clinker and the second is the reduction of clinker to a powdery state with the addition of gypsum or other additives [1].

The production of cement, the composition and properties of which correspond to modern Portland cement, practically began in the first half of the XIX century in England (since 1825). The first cements were produced in batch shaft kilns, mixing the components in mechanical mixing plants. It was only after 1900 that shaft kilns were replaced by more efficient continuous rotary kilns. Therefore, comparison of technologies by technological cycles of production development is possible only from the beginning of the 20th century, when cement production was modernised and wet, dry and combined production methods appeared.

In Russia, natural limestone, chalk, marl and other carbonate-containing materials, including wastes from related industries, are used as the main raw materials for cement clinker production. As a clay component natural clays, overburden clay materials, clay shale, slags, ash and slag wastes, etc. are used. (in total more than 40 items). To adjust the raw mix, a number of man-made iron-containing materials are used - sinter cinders, dust and sludge from gas cleaning facilities of ferrous metallurgy, converter slag, sludge briquettes and other iron-containing materials. The share of natural resources in the raw mix for clinker production is about 70%.

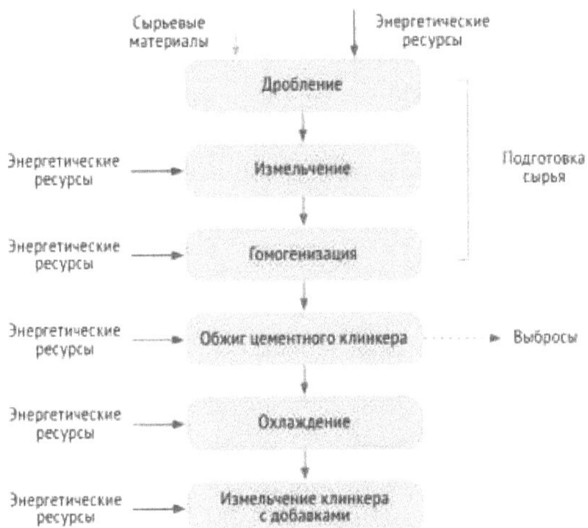

Fig. 5. General scheme of the technological process of cement production [1].

Raw materials in cement production are mainly clay and carbonate rocks, as well as other natural raw materials and some types of industrial wastes, slags, etc. [1,2,3].

With the development of technology in construction, people started to think more and more about possible negative effects on the environment. Concrete, particularly Portland cement, is one of the most well-known materials for construction due to its low cost and ease of use [4]. The main component of concrete is cement, which is the main threat of greenhouse gas emissions [5]. The domestic production technology is the production of cement in a "wet process" in which the virgin material, called slurry, is a substance with a moisture content of up to 40 per cent [3]. Then there are several main steps: adjustment, firing, addition of gypsum and other additives and grinding. It is worth noting that this production method requires more fuel for the firing stage than the "dry method". There is evaporation of water from the sludge [2].

Carbonate rocks are limestone, limestone-shell, chalk, marly limestone, marl, marl, metamorphic or sedimentary rocks of dolomitic, carbonate-clay and limestone composition. The quality and value of such rocks as raw materials for cement production is determined by their structure and physical properties. Rocks with crystalline structure interact with other elements of the mixture during firing worse than rocks with amorphous structure. Chalk is an easily grindable soft sedimentary rock, a type of smeary limestone. It is easily pulverised and is a popular raw material for creating cement.

Marl is a sedimentary rock, transitional from limestone to clay. It can have a hard or loose structure, different density and humidity depending on the percentage of clay admixtures. Building mortars based on marl are actively used in the construction of cookers, fireplaces, etc.

Among limestones, porous and marly types with a low compressive strength threshold and without silicon inclusions are preferred for cement production. Clayey rocks are used in cement production: loam, clay, loess, shale and loess-like loams.

Fig.6. Manufacturing of raw materials for cement production [3].

Clays, sedimentary rocks, composed of various kinds of minerals, become plastic and swell when moistened. In the dry process of cement production, the binding capacity and

The plasticity of clay makes it possible to granulate flour and briquetting. A loam is a clay that contains a high proportion of dusty and sandy particles.

Clay shale is a dense and hard rock that can easily stratify into laminae of small thickness. Relative to clay, shale has a more constant composition and lower moisture content.

Loess is a fine-grained, loose and porous rock composed of fine particles of clay, feldspar, quartz and other silicates. Loess is not characterised by high plasticity. Loess-like loam is a material with properties transitional between loam and loess [3].

In addition to the basic raw materials, various types of corrective cement additives are actively used in the production process to change some properties of the final product. These can be alumina, silica, clay-containing additives, as well as fluorspar as mineralisers (sodium silicofluoride, gypsum, apatite, phosphogypsum, fluorite)

It should be noted that the raw material composition of both dry and wet cement production methods can vary depending on the location of the cement plant, the availability of certain raw materials, equipment capabilities, the demand for certain types of products in the region and much more [3]. The main component of cement production   is clinker   .

An intermediate semi-finished product obtained by firing a mixture of 75% limestone (chalk, marl or other rocks) and 25% clay. The raw materials are melted to form granules. The clinker is ground and combined with ground additives.

The whole process of cement binder production can be divided into 3 stages:

• clinker production by firing is the main process, the most costly and labour-intensive;

• grinding clinker to a fine powder;

• mixing clinker powder with powdered additives.

The manufacture of clinker is divided into the following stages:

• delivery of clinker raw materials to the cement plant;

• grinding of raw materials;

• mixing the components in the proportions specified in the technical documentation for subsequent firing.

There are several technologies for the production of cement.

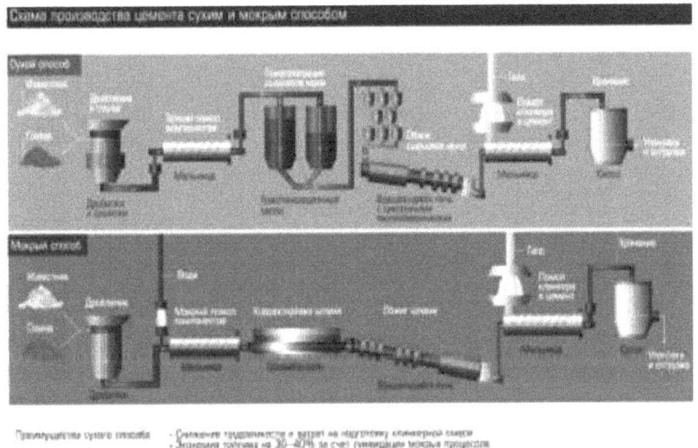

Figure 7. Cement production technology [4].

The comparison of cement production technologies shows. Dry and combined production methods are considered as energy saving methods. Each of the above mentioned cement production methods (wet, dry, combined) has its own advantages and disadvantages.

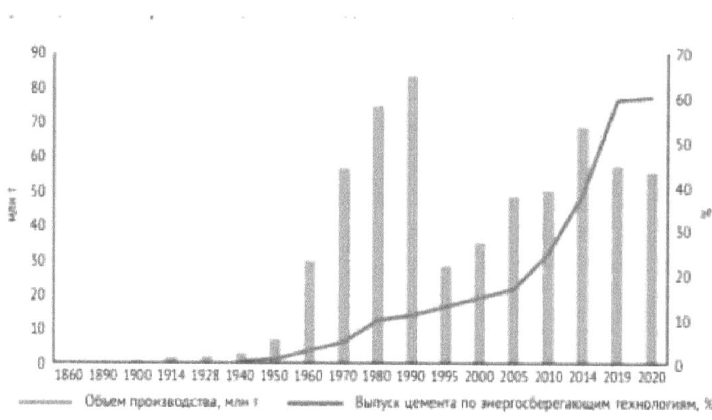

Figure 8. Comparison of technologies by technological cycles by key indicators

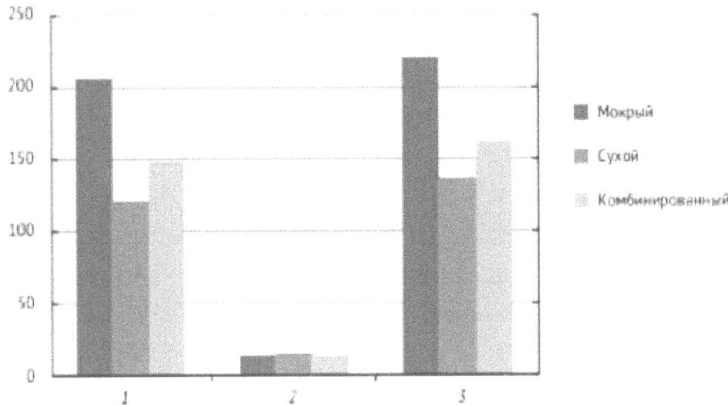

Figure 9. Annual averages of energy resources by method production, kg c.t. (1 kg = 8.141 kWh): 1 - specific fuel consumption; 2

- specific energy consumption; 3 - reduced energy costs

So, for example, in the presence of water facilitates the grinding of materials and easier to achieve homogeneity of the mixture, but the heat consumption for firing the raw material mixture in the wet method is 30 -40 % more than in the dry method (Fig. 9).

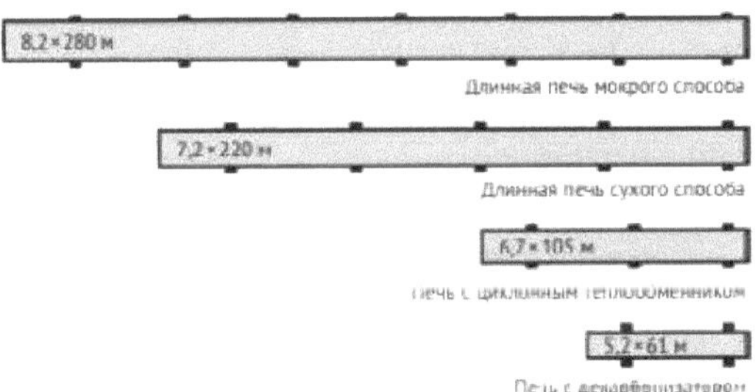

Figure 10. Calculated dimensions of a rotary kiln for the production of 5000 tonnes of clinker per day (according to ZKG 4/2019)

The choice of production methods of Portland cement clinker is determined by a number of factors of technological and technical-economic character: properties of raw materials, their homogeneity and humidity, availability of sufficient fuel base in the area of plant construction. At natural humidity of raw materials more than 18 -20 % the wet method is expedient. This method is also favourable when using two soft components (clay and chalk), as their grinding is easily achieved by stirring in water.

The wet method is used in case of variegated chemical composition. The dry method is rationally applied in case of homogeneous raw materials, if their humidity does not exceed 18-20 %. In practice there are examples of successful functioning of enterprises using chalk and marl with humidity up to 26 % (Belarusian cement plant, built in 1980s, has been working on such raw materials since the end of the last century). The semi-dry method will give good results when clinker is produced from sufficiently plastic raw materials, when strong and heat-resistant granules are formed when the mixture is granulated. In case of good filterability of raw material slurries the combined method should be favoured. In the dry method of production, limestone and clay after leaving the crusher are dried to a moisture content of about 1 % and ground into raw meal. After grinding it is dosed, averaged and adjusted in special mixing silos and fed to cyclone heat exchangers [1].

The main advantages of the dry method of Portland cement clinker production are: - higher clinker removal from 1m2 of kiln unit than in the wet method; - economical method (reduction of fuel consumption, energy costs, cost of 1 tonne of cement). At the wet method the power consumption for grinding is lower, as the grinding of materials in aqueous environment is easier, and some raw materials are able to soften (dissolve) in water - clay, chalk. If both raw materials are soft (chalk + clay), energy saving during grinding can reach 10 kW^h per tonne of raw materials. Averaging of the charge is easier and more reliable compared to the powdered state in the dry method. At preparation of raw material sludge it is necessary to introduce additionally from 30 to 50 % of water, and then it is necessary to remove it by evaporation (in cement kiln), and it is connected with excessive expenses of fuel and energy - in 1,5 -2,0 times more, than at dry method.

As a result, at wet method specific heat consumption varies from 5800 to 6500 kJ/kg of clinker, and at dry method it is 3100-3500 kJ/kg of clinker, which leads to reduction of production cost. At dry method of charge preparation drying of raw materials is carried out: before grinding or simultaneously with grinding in crushers or mills. At wet method of production sludge is moved by hydraulic transport - by gravity or with the help of centrifugal pumps, at dry method pneumatic transport, screws and elevators are used, which increases dust pollution of air in shops and on the territory of the plant and requires installation of additional equipment for dedusting of aspiration air [1].

The volume of kiln gases in the dry process is 35-40 % less than in the wet process at the same kiln capacity. As a result, the dry method of production reduces the cost of dedusting kiln gases, there are more opportunities to use the heat of kiln exhaust gases for drying raw materials, which reduces the total fuel consumption for clinker production, although it causes complication of production technology. Thus, newly constructed and designed cement plants in our country are designed to produce cement by the dry method as a more economical method (Fig. 34). The same trend is observed in all industrially developed countries of the world. The average fuel consumption in Russia for firing a tonne of clinker in the wet method of production is 206.5 kg c.t. against 121.5 kg c.t./tonne of clinker in the dry method of production. Approximately 115 -125 kWh of electricity is consumed to produce the same amount of cement. In industrially developed countries these figures are one and a half times less, and the output per cubic metre of kilns in Russia is significantly lower. Fuel and electricity costs account for an average of 40 per cent of the cost of our cement, and at some enterprises even 60 per cent.

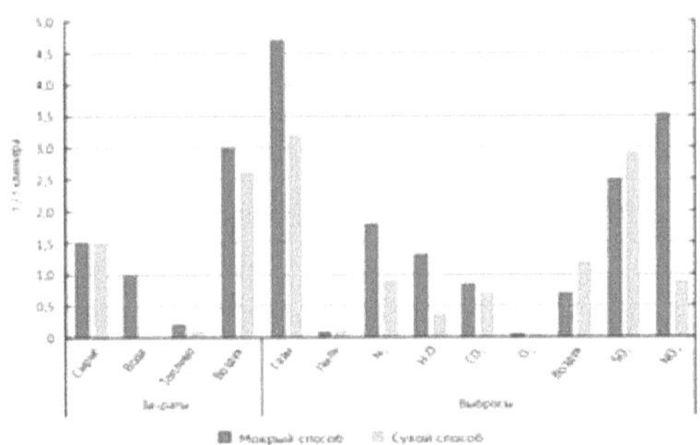

Figure 11: Material balance of the process of production of ortland cement clinker by wet and dry production methods [1].

From an environmental point of view, the wet process has the greatest negative impact on the environment, taking into account material inputs and pollutant emissions (Fig. 10).

The consumption of material resources and emissions increases by 1.5 tonnes per tonne of clinker. Production of Portland cement clinker by the wet method requires approximately twice as much fuel as by the dry method. The heat generated by the combustion of process fuel is used for the thermal processes of clinker formation, water evaporation, as well as lost with waste gases, with air from the clinker cooler, with hot clinker and for direct losses to the environment. On the other hand, emissions of hot tertiary air are slightly increased in the dry process due to the use of modern clinker coolers,
cooling Portland cement clinker to temperatures below 80 -100^0 C. Aerosols from rotary kilns of the dry method are characterised by high dispersion of particles (up to 75 % of particles smaller than 5 microns) and high temperature in comparison with the wet method. But due to the reduction of material consumption, fuel consumption, gas and dust emissions into the atmosphere, this technology is more progressive both in economic and ecological aspects.

Efficiency of thermal processes, as a rule, is characterised by the coefficient of efficiency (COE), representing the ratio of useful energy to the energy used. However, in a number of cases, the energy (thermal) efficiency characterises the process insufficiently. In the 1950s, a method based on the combined use of the laws of thermodynamics, taking into account the special role of the environment, was used to evaluate the efficiency of thermal engineering processes; the term "exergy" appeared.

By exergy is meant the work that can be obtained from a system during its reversible transition to a state of equilibrium with the environment. Thus, exergy is the part of energy contained in a system that can be utilised from a practical point of view. The exergy is determined by the difference in the values of the properties of the system and the environment. The greater this difference, the higher the exergy is, and if the system has environmental parameters, its exergy is zero. The value of exergy depends on the environmental parameters (temperature, pressure and composition). The exergetic method is an effective way to analyse and optimise thermal process systems and allows to evaluate the degree of perfection of heat and energy transfer and conversion processes. Exergy analysis is divided into two types: the method of exergy losses, or entropic method, which consists in calculation of exergy losses in all individual typical processes that make up the system under consideration, and the method of exergy flows, or balance method, which consists in calculation of exergy of input and output flows of the system and compilation of exergy balance and determination of exergy losses as a difference of exergy of input and output flows.

Theoretical methods of exergetic analysis can be considered to be developed quite completely, but in applied problems it is used mainly in the study and optimisation of low-temperature heat exchange processes. In industrial furnaces a useful product is produced from non-energetic raw materials, and energy for this purpose is supplied by combustion of fuel. In addition to the useful products, tailing products are also produced, which are not used, but their formation is an integral part of the technological process.

In cement kiln systems, heat transfer processes are auxiliary to the main task of carrying out structural and chemical transformations in the initial product in accordance with the required temperature regime. Thus, the possibility of transferring exergetic methods to clinker kilns remains questionable and requires further study. In 1996-2005. M. E. Verdijan and his co-workers carried out a new approach to the exergetic analysis of the cement production process. They defined the exergy of the transformed material as its ability to respond to the impact exerted on it. The exergy value of the initial charge is determined by the exergy of the selected components. It follows that it is necessary to endeavour to select initial components having high values of their exergy. It is shown that the exergy of the charge E of the charge is dependent on the parameters characterising the raw components used, the technology of their preparation and firing:

$$E_{шихты} = f(KH, R008, Pe, W, T, S),$$

where KN - chemical composition of the raw material charge; R008 - degree of grinding of the charge; Pe - degree of mixing of the charge; W - moisture content; T - degree of heat treatment; S - technology of preparation and firing of the charge. The performed calculations have shown that the maximum exergy of raw materials takes place at its minimum humidity and maximum porosity. Therefore, by selecting the appropriate raw materials, the minimum energy consumption can be laid already at the first stage of cement production. Natural raw materials and dry products of fuel combustion in the proposed environmental model are referred to reference substances, their chemical exergy is equal to zero. Therefore, the chemical exergy in the clinker firing process will be non-zero for fuel, clinker, water vapour and man-made products used as raw material additives. However, other researchers, commenting on the results obtained, pointed out that "a serious drawback of the discussed material seems to be the lack of indications on the way of calculating exergy" and that the values given by M.E. Verdiyan are not exergy in the traditionally understood sense, but a part of the expended energy required to obtain the finished product32. In this case, the considered approach can be fully considered as an energy analysis of cement production technology. At the same time, the works by M. E. Verdian, P. V. Besedin and P. A. Trubaev consider separately the exergy analysis of cement clinker firing processes and the calculation and formation of cement exergy in mills. Let us compare the energy efficiency of different technological processes of clinker roasting in rotary kilns with grate clinker coolers.

Thermal efficiency, characterising heat losses in the process, for the considered variants varies little, and its worst value is for the dry method with the lowest fuel consumption. This is explained by the greatest losses in this method of heat with waste gases and through the furnace body. But the exergy efficiencies of PE and PE.p increase with decreasing fuel consumption, which is due to a decrease in the exergy of fuel at the inlet with a constant exergy of clinker and little changing exergy of waste gases at the outlet. Since the heat input for chemical reactions of raw material conversion into clinker is constant, the process efficiency increases with decreasing fuel consumption.

Cement industry is a large ecological reserve with all possibilities for effective utilisation of various secondary raw materials: gas, metal, oil, coal, biomass, car tyres, agricultural waste and household waste (Table 1).

Table 2. Production and utilisation of anthropogenic materials

Name of waste	D ONE way out	Utilisation, %	Availability in dumps
Ash and slag from thermal power plants. mln tonnes	40	7	1500
Blast furnace slag, mln tonnes	40	20	360
Non-ferrous metallurgy slags, mln tonnes	so	2	450
Electrothermophosphate slag, mln tonnes	20	8	130
Steelmaking slags, mln tonnes	10	0,5	70
Wood waste, mln m*	up to 100	up to 20	n/a
Phosphogypsum, mln tonnes	25	5	280
Pyrite cinders, mln tonnes	7	30	40
Limestone sands, mln tonnes	175	7	n/a
Worn tyres, mln tonnes	12	4	n/a
Construction and demolition waste, mln m⁵	13	20	n/a
Municipal solid waste, mln m	180	3	n/a

Name of waste Annual yield Utilisation, % Presence in dumps Ash and slag from TPPs, mln tonnes 40 7 1500 Blast furnace slag, mln tonnes 40   20360 Non-ferrous metallurgy slag, mln tonnes   502450 Electrothermophosphate slags, mln t 20 8 130 Steelmaking slags, mln t 10 0.5 70 Wood waste, mln m3 up to 100 up to 20 n/a Phosphogypsum, mln t 25 5 280 Pyrite cinders, mln t 7 30 40 Limestone sands, mln t 175 7 n/a Worn tyres, mln t 12 4 n/a Construction and demolition waste, mln m3 13 20 n/a Municipal solid waste, mln m3 180 3 n/a

Since different types of waste materials can be used as raw materials, the basic principles of their use, such as pre-sorting and analysing the processes for their preparation, should be considered before deciding on their use. Portland cement clinker is characterised by a specific composition that predetermines the hydraulic properties of cement. This means that all raw materials and fuel ash must be carefully matched in terms of mineral composition and feed rate to obtain the specified clinker composition. In order to maintain a standard clinker quality, preliminary studies on the effect of waste on clinker formation processes must be carried out. The final decision on what type of waste will be accepted for use in a particular plant cannot be generic. Secondary resources replacing carbonate and clay components include blast furnace and non-ferrous metallurgy slags, ash and slag waste, nepheline sludge, gravel and aggregate production cuttings, and the following

In the meantime, it is necessary to use less capital-intensive and efficient technologies that have existed for a long time, such as grinding raw materials and cement in a closed cycle. In this case, the quality of cement is much higher and electricity consumption for its grinding is reduced by 15-30 per cent. There are a number of other tried and tested technical solutions that ensure a significant reduction in fuel consumption for clinker burning in the wet method of clinker production. One of them is the reduction of sludge moisture through the use of new, more efficient sludge thinners, which allow to reduce the specific fuel consumption up to 15-20 kg. The use of new high-performance materials for lining the preparation zones of rotary kilns also helps to save resources, while additional feeding of kilns with technogenic products increases their productivity and improves their efficiency.

reduces fuel consumption by at least 10 %. All the above-mentioned allows to bring the wet method of production closer to the dry method in terms of specific consumption indicators, to ensure its competitiveness. Introduction of modern equipment with cyclone heat exchangers and decarbonisers, with fifth-generation refrigerators at the most modern grinding plants with predisgrinders and separators of the latest models will allow consuming no more than 100-110 kg of fuel per tonne of clinker. In order to reduce energy consumption and improve energy efficiency at enterprises, measures should start with the implementation of primary technical solutions integrated into the technological process. Such primary measures may include the following: - optimisation of firing and cooling processes of Portland cement clinker; - improvement of homogeneity of raw material composition; - improvement of accuracy of fuel dosage; - regulation of gas-dynamic mode of kiln and cooler operation; - use of computer systems of unit control. Considering the principles of energy efficiency in relation to cement production, the following ways of reducing energy costs can be emphasised: - use of "non-traditional" raw material components; - use of alternative fuel; - utilisation of waste gas heat for electric power generation; - reduction of clinker share in cement; - use of new promising types of cement, such as solidia cement, celitement and biocement, geopolymer and belite cement; - technological progress and innovations. In 2015, SM Pro LLC made a forecast of the economy of cement enterprises in Russia - calculated the cost of producing 1 tonne of cement by dry and wet production methods, taking into account the cost of attracted [1].

In the meantime, it is necessary to use less capital-intensive and efficient technologies that have existed for a long time, such as grinding raw materials and cement in a closed cycle. In this case, the quality of cement is much higher and electricity consumption for its grinding is reduced by 15-30%. There are a number of other tried and tested technical solutions, which provide a significant reduction of fuel consumption for clinker firing in the wet method of its production. One of them is the reduction of sludge moisture through the use of new, more efficient sludge thinners, which allow to reduce the specific fuel consumption up to 15-20 kg. The use of new high-efficient materials for lining the preparatory zones of rotary kilns also helps to save resources, and additional feeding of kilns with man-made products increases their productivity and reduces fuel consumption by at least 10 %. All the above-mentioned allows to bring the wet method of production closer to the dry method in terms of specific consumption indicators and to ensure its competitiveness. Introduction of modern equipment with cyclone heat exchangers and decarbonisers, with fifth-generation refrigerators at the most modern grinding plants with predisgrinders and separators of the latest models will allow consuming no more than 100 -110 kg of fuel per tonne of clinker. In order to reduce energy consumption and improve energy efficiency at enterprises, it is necessary to start measures with the implementation of primary technical solutions integrated into the technological process. Such primary measures include the following: - optimisation of the processes of firing and cooling of Portland cement clinker; - improvement of homogeneity of raw material composition; - improvement of the accuracy of fuel dosage; - regulation of the gas-dynamic mode of operation of the kiln and cooler;

- use of computerised systems for controlling aggregates. Considering the principles of energy efficiency as applied to cement production, the following ways of reducing energy costs can be emphasised: - use of "non-traditional" raw material components; - use of alternative fuel; - utilisation of waste gas heat for electric power generation; - reduction of clinker share in cement; - use of new promising types of cement, such as solidia cement, celitement and biocement, geopolymer and belite cement; - technological progress and innovations. In 2015, SM Pro LLC made a forecast of the economics of cement enterprises in Russia - it calculated the cost of producing 1 tonne of cement by dry and wet production methods, taking into account the cost of attracted cement.

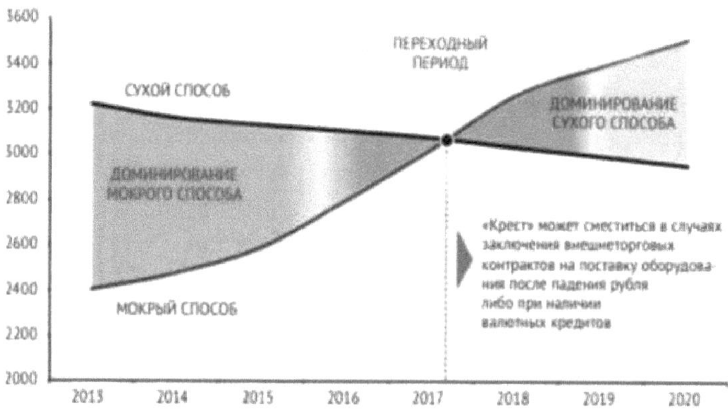

Figure 12: Costs of producing 1 tonne of cement by dry and wet process

methods in 2013-2020, rub./t (actual and forecast) [1].

According to Evgeny Vysotsky, Executive Director of SM Pro LLC, Russia has enough capacity to meet the needs of the construction industry until 2025. Established

the market surplus is the result of two components: on the one hand, the lack of demand (and here the government's plans for the construction industry give hope for a possible change in the future), and on the other hand, the introduction of new lines with a total capacity, taking into account the current range of products, of about 35 million tonnes of cement per year. Speaking in September 2019 at the XVII International Central Asian Cement Industry and Market Conference Business Cem Tashkent 2019, E. Vysotsky, having conducted a strategic analysis of the competitiveness of each enterprise, showed that in the current conditions the Competitiveness Index (K -Index) of individual new dry enterprises is comparable to the K -Index of individual wet enterprises. According to E. Vysotsky, this means that in addition to the method of production, the competitiveness of the enterprise is influenced by the remoteness of the enterprise from the key sales markets, i.e. logistics. At present, the choice of production method is not a determining factor in assessing the competitiveness of an enterprise that guarantees a stable financial result to the owner. Dry or wet production method is a question of production cost. However, apart from the production cost, financial costs per tonne and logistics are essential components of the cost of goods to the consumer. At the same time, it is necessary to remember about the environmental friendliness of production. Today, at wet process plants with modern bag filters or electrostatic precipitators, emissions (especially of suspended solids) are significantly lower than at some dry process plants commissioned in the last 10-15 years. Thus, today, dry or wet production is not a key factor of competitive advantage. The clinker factor, or clinker/cement ratio (i.e. clinker content in cement), is indicative of the environmental friendliness of cement production: the lower the ratio, the less greenhouse gas CO_2 is emitted when 1 kg of cement is produced (Figure 45)

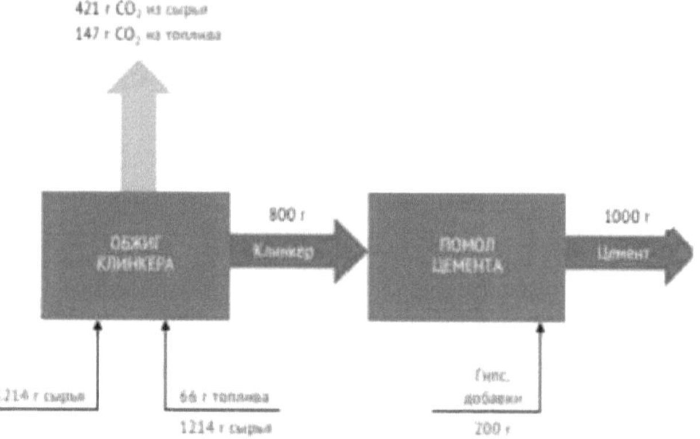

Figure 13. CO2 emissions during the production of 1kg of cement.

Greenhouse gas emissions, especially carbon dioxide (CO_2), are mainly related to fuel combustion and to the decarbonisation of limestone, which in its pure form contains 44 wt% CO_2. Cement production has been prioritised as a priority source of industrial greenhouse gas emissions not only in the EU but also in many countries around the world. However, carbon dioxide is not included in the list of normalised production parameters. On the contrary, in the European Union, CO2 emissions from cement production are included in the Greenhouse Gas Emissions Trading System. In many countries around the world, sectoral guidelines for improving energy efficiency and, in some cases, for reducing greenhouse gas emissions are being developed and implemented. Sectoral guidelines for improving energy efficiency

energy efficiency are Energy Star (USA), Energy Management Guidelines (UK), Industrial Emissions Directive (EU).

The Cement Sustainability Initiative (CSI), together with the International Energy Agency (IEA), has set emissions of 370 kg CO2 per 1 kg of Portland cement as an acceptable target for 2050. This target corresponds to a scenario of a 2^0 C increase in global ambient temperature by 2100. In addition to proper cement clinker firing operations, the following methods of CO_2 prevention and control are recommended: - selection of a process and operating mode that promotes energy efficiency   (drying /

preheating / pre-combustion); - selection of fuels with low carbon to calorific content (e.g. natural gas, diesel or some waste used as fuel); - increased use of biogenic (CO_2 neutral) fuels; - selection of raw materials with low organic matter content; - production of multi-component cements, which have the potential to significantly reduce fuel consumption and, consequently, CO2 emissions per tonne of final product. Globally, the clinker factor has been declining since

1990. Analysis of the data presented shows that the lowest clinker/cement ratio is characteristic of India and Brazil, which are the leaders in cement production. Therefore, the production of multi-component cements (where the clinker-factor is 0.70 - 0.75) allows to significantly reduce the consumption of both raw materials and fuel, which, accordingly, leads to a reduction in CO_2 emissions. In this regard, the type structure of cement production is given a lot of attention. In the type structure of cement production, it is worth noting that the share of Portland slag cement production has been decreasing for a long time. If in 1990 its share in the total output was about 28 per cent, at present it is about 3 per cent. This leads to an increase in the cost of cement and, consequently, to an increase in its price. In 2010, the industry produced 48 per cent of unblended cements and 52 per cent of blended cements, and in 2020. The output of blended cements decreased to 31 %, while the share of Portland cement without mineral additives increased by 12 % to 34.95 million tonnes. Our plants produced 19.77 million tonnes of Portland cement with additives, which is 1.5% less than in 2019. Production of Portland slag cement (2.5% of the total cement production in the country) decreased by 2.9% to 1.4 million tonnes. There has been a steady decline in blended cements. In addition to slag, clinker can be replaced by fly ash, pozzolans, limestone, etc. [1].

To increase the efficiency of grinding processes in cement technology, physical and chemical methods of intensification are used, based on the creation of adsorption-active medium in the mill by introducing small amounts of surface-active substances (surfactants) into the first chamber in a fine-dispersed (atomised) state. Amino-alcohols, higher alcohols, lignosulphonates or their compositions and a number of other additives are considered to be effective intensifiers of grinding in modern production of binding materials. Widely used nowadays intensifier of grinding process technical triethanolamine (TEA) - $N(CH_2-CH_2-CH_2)_3$ is a scarce and expensive product, in this connection its introduction is limited, that caused the necessity to find new highly effective, cheaper surfactants. Moreover, in the synthesis of surfactants it is very promising to use wastes of various industries as feedstock, which contributes to reducing the cost of additives and utilisation of industrial wastes and formed the basis of this study.

1.2 Ways to reduce energy consumption in cement industry.

Ecological problem and application of industrial wastes in cement production. In cement production various wastes from other industries are widely used, the expediency of their application is dictated by two main factors: the need to protect the environment (improving the ecological situation in different regions of the country) and the desire to achieve high technical and economic performance, mainly reducing fuel and energy costs.

The utilisation of industrial waste is carried out in three main directions: introduction into the raw material charge during clinker firing; introduction into the cement composition as a setting time regulator, intensifier of raw material and clinker grinding.

The use of secondary resources in cement production allows to reduce the use of natural raw materials and additives in the cement industry as a whole by 18, 8%. At present more than 70 million tonnes of secondary raw materials are utilised in the cement industry [6-9], including 30.7 million tonnes of blast furnace granulated slags, 5.2 million tonnes of nepheline sludge, 3 million tonnes of iron additives, 2.7 million tonnes of electrothermophosphorus slags, 1 million tonnes of fuel slags and ashes.

The use of waste is becoming increasingly necessary for many cement plants due to the depletion of natural raw materials and the need to intensify and economise cement production. Clay quarries are being depleted particularly quickly, as their bed capacity is usually small. Explored new deposits are usually far from the plant and often under agricultural land.

The use of waste in the cement industry allows solving the problem of waste utilisation most radically, firstly, from the quantitative side, since cement production is mass and multi-tonnage, and cement plants are located in almost every region. Secondly, the use of industrial wastes in cement production has a high technical and economic efficiency, primarily due to the integrated use of raw materials and reduction of land allocation for quarries and dumps. Besides, the use of wastes, which have already undergone thermal treatment, always leads to reduction of fuel consumption for cement production. Thirdly, this issue is important for the development of the cement industry itself in terms of securing raw material base and improving the quality of cement. Fourthly, cement production is waste-free and allows not only efficient utilisation of wastes, but also more rational and comprehensive use of natural materials, e.g. cement industry is the only field where all components of solid fuels are fully utilised. The combustibles burn and provide the heat required for the process, while the ash is fully incorporated into the finished product - clinker, reducing the consumption of the natural component clay. In all other industries, ash from hard coal combustion is a waste product.

Thus, the issue of waste utilisation in the cement industry can and should be radically solved at the stage of design and construction of new waste suppliers and cement plants. At the same time, under specific conditions it is possible to effectively utilise waste at existing cement plants. In this case, there are often problems that need to be solved [5-7J. Firstly, it is the processability of the material in production and during transport. It should be noted that the majority of waste coming out directly in production without additional processing is not technological.

Secondly, there are transport issues, as the waste is located in one place and the cement plants in another. Depending on the distance and the availability and condition of roads, both road and rail transport can be used, which needs to be taken into account when assessing the economic efficiency of waste utilisation.

Thirdly, to ensure the stability of the technology and cement quality, the homogeneity of the chemical composition of the waste used over time is of great importance. In most cases, the waste is obtained from enterprises whose production technology is not less stable than cement technology, and therefore the homogeneity of the chemical composition of waste is quite satisfactory for cement production.

Fourthly, in a number of cases it may be necessary to change the technology of cement production due to the use of waste materials with very different properties from natural raw materials. In this case scientific research and development of a new technology will be required.

Fifth, with the widest use of waste, the cement industry, which produces mainly Portland cement clinker-based cements, cannot consume large quantities of some wastes because their consumption is limited by the silicate composition of the clinker.

The huge volume of cement production in the country requires significant consumption of fuel and electricity. In the cost of cement production the share of fuel is 25%, electricity - 14%, and in the cost of clinker the fuel costs are up to 43% [8]. Reduction of fuel and electricity costs in the cement industry is of great importance, as it allows to reduce the cost of production and save significant amounts of fuel and electricity, which is of great national economic importance.

One of the important ways to solve this problem is the wide use of various chemical additives in the cement industry, A very small amount of these additives can reduce energy costs and intensify the technological processes of cement production, as well as accelerate the hardening of the binder and reduce its consumption in concrete and mortars and improve the quality of the latter.

Global powder production now reaches 23 billion tonnes per year. which consumes about all the electricity generated.

Grinding of materials in cement technology is also the most energy-intensive process. Grinding of materials has a decisive influence on the quality of produced clinkers and cement, as increasing the degree of grinding of raw materials up to certain limits improves their sinterability, improves the quality of clinker and ensures its uniform composition [9,10].

Crushing and grinding operations consume 70-80% of the electricity used for cement production. When using hard rock carbonate raw materials and solid fuel, electricity consumption per 1 tonne of cement is 100 -120 kWh, when working on soft raw materials and liquid or gaseous fuel - 60 -80 kWh.

One of the measures allowing to reduce power consumption for raw material and clinker grinding process is the use of intensifikators of grinding - surface-active substances (surfactants).

Due to the increasing share of dry cement production, in recent years, work has also been carried out on the intensification of dry grinding of raw materials [11]. The introduction of surfactants in dry grinding of raw materials will reduce the consumption of electricity, compressed air and intensify the operations of averaging, homogenisation, transportation and unloading of raw meal [12,13].

Depending on the hardness of the materials to be ground, the fineness of grinding and the type of grinding unit, the clinker grinding process consumes up to 40 per cent of the total energy balance in cement production, or 35--40 kWh.

The analysis of literature on energy consumption reduction shows that the issues of energy consumption reduction, increase of mill productivity and production of fast hardening high-strength cements with high specific surface area require mainly, apart from other measures, wide application of cement grinding intensifiers. Such surfactants as triethanolamine (TEA), monoethanolamine (MEA), sulphite-alcoholic bard (SAB), soapstock, soap-oil, technical lignosulphonates and others have been researched and found application, allowing to increase the productivity of mills by 15-25% and reduce energy consumption up to 15% [14-17].

Chapter 2

Cement clinker.

2.1 Peculiarities of the process of grinding of solid materials in a surface-active medium

Cement clinker is a polycrystalline elastic-plastic solid consisting of 75-82% silicate minerals and 18-25% fused minerals. Minerals: clinker have different brittleness. The most brittle were minerals: intermediate and alite, the least brittle is belite [18].

The difference in brittleness is explained by the large defectiveness of the crystal structure of the minerals and the presence of "voids" due to the irregularity of ionic coordination [19].

Cement is ground in three stages [18], in the first stage the work of grinding up to 1200 -1500 cm^2 /g is proportional to the newly formed surface of the ground material, clinker is destroyed by weak points, defects in the structure . Grinding resistance in the interval from 1200 - 1500 to 2300 -2700 cm^2 /g increases sharply at the second stage. Grindability at this stage depends on the microstructure of clinker: size, shape, nature of crystal aggregation, quantitative content of glass phase. When grinding clinker to a specific surface of 2300 - 2700 cm^2 /g and more (III stage), the rectilinear dependence inherent in the first two stages is broken, which is caused by the phenomena of sticking and aggregation.

At fine grinding of clinker, the smallest particles of the ground cement adhere to the grinding bodies and internal surfaces of the mills with a rather strong layer, and also aggregate with each other with the formation of lumps, flakes, plates. These phenomena sharply worsen the conditions of the grinding process energy consumption for additional crushing of newly aggregated cement particles, for plastic irreversible deformations of the adhered layer, increase in the amount of friction work reduce clinker.

Reducing the impact energy of the grinding bodies reduces the productivity of the mills by 20 -25% and increases the temperature inside the mills. Further grinding intensifies sticking and aggregation, which leads to cessation of specific surface area growth or even to its reduction. All energy spent on grinding goes to the formation of aggregates [9,10, -20 -13].

It is believed [24] that the process of sticking occurs due to pressing of cement particles into defects in the surface of grinding bodies and armour lining. Other authors [25-27] explain the phenomena of sticking and aggregation by the formation of uncompensated molecular forces on the surface of cement particles during dispersion, adsorbing vapour and gas molecules from the environment.

The electrostatic theory of the electric double layer is widely spread [28-30].

According to this theory, during clinker grinding the phenomenon of contact electrification occurs due to the rupture of the double electric layer at the boundary of the cement particle and the environment [28]. In this case, cement particles are electrically charged, oppositely charged particles are attracted and their aggregation occurs, the surface of grinding bodies during grinding acquires an opposite charge to the cement particles, which leads to the phenomenon of sticking.

The fracture process occurs under the application of mechanical forces, which contribute to the expansion of existing microcracks and the formation of new nucleated cracks in the layers of a solid body. The processes of nucleated cracks formation in cases of brittle fracture of bodies are well explained by the dislocation theory, according to which low-temperature fracture of bodies is preceded by plastic deformation localised as a result of slip or twinning, which is inevitable even in cases of rapid crack development [31].

Cracks nucleate as a result of the formation of dislocation clusters and their fusion in front of twin boundaries, block and grain boundaries or surface films [32]. Grain size also affects the magnitude of the fracture resistance of a solid. As the grain size increases, the length of the continuous slip path increases, which increases the probability of crack initiation [31,33]. On the contrary, decreasing the grain size increases the resistance to brittle fracture, because an increase in the number of boundaries between grains will increase the energy absorption processes and thus hinder the development of cracks.

Fine grinding in the cement industry is carried out in ball mills. Under the action of the crushing load the pieces of material manifest themselves into fine powder with high free surface energy.

Surface interaction of dispersed material particles has a negative impact on the efficiency of the ball grinding process, as it leads to sticking of the crushed material on the working surfaces of the crushing, loading and armour plates. This reduces the efficiency of deforming force transmission from the working bodies to the crushed material.

The most effective way to reduce the surface interaction of dispersed material particles is the addition of surface-active substances (surfactants). These substances, adsorbing on the surface of the particles of the material reduce its surface energy, as well as the number and area of possible surface contacts of the solid phase. In addition, surfactant molecules, covering the surface of material particles, contribute to its destruction. This phenomenon was discovered in 1928 by P. A. Rebinder and is called adsorption reduction of solid body strength [34.

According to the theory of P.A.Rebinder and B.V.Deryagin [35,36], the reduction of strength or hardness of a material in the process of its dispersion is caused by the fact that a solid body contains many inhomogeneities, starting from defects in the crystal structure and ending with microcracks of different sizes in a piece of material. Under the influence of external mechanical forces, a zone of increased fracturing, the so-called pre-fracture zone, appears in the solid body.

Opened microcracks, by virtue of mutual attraction due to the creation of excess free surface energy on their walls, after removing the load are again closely closed. Surfactants, adsorbed on the surface of crushed materials penetrate into microcracks, realise dense layers lowering the force of attraction of solid particles and prevent the closure of microcracks, as well as produce wedging action, which significantly reduces the work expended on the destruction of a solid body. This fact is of particular importance in case of repeated application of forces in the process of mechanical fracture.

The analysis of the existing methods in various fields of chemical technology for assessing the surface interaction of disperse materials shows that the surface interaction of particles of solid materials is most appropriately assessed by indicators characterising the process of their pressing [37]. The process of pressing bulk materials is usually divided into three main stages.

At the first stage, the squeezing force is spent to overcome the cohesive forces acting between the particles of dispersed material and contributes to its significant compaction. At the second stage of pressing, local contact plastic deformation of the material occurs, accompanied by a slight compaction. The third stage is characterised by deformation in the whole volume of particles and is accompanied by their destruction. Based on these fundamentals of the theory of pressing of powdery materials, we can assume that in a ball mill dispersed material is subjected to three stages of pressing when a metal ball falls on it. In this case, part of the work done by the falling ball is spent on overcoming the forces of surface interactions of particles associated with the presence of free energy on their surface (easily destructible and recoverable surface contacts). Part of the work of the falling ball is spent on the creation of strong surface contacts of particles as a result of their plastic deformation. Since the sum of these works performed by the falling ball predetermines, mainly, the additional costs of grinding associated with surface interactions of particles of the materials being ground, the value proportional to it can serve as a quantitative parameter characterising these surface interactions.

Studies have established [11] that such a value is the work A_1 , spent on pressing I gramme of material to create strong and fracture easily breakable and recoverable contacts acting on the unit of its contact surface:

$$A = P\,K_1(W_0^{\,1} - W_1)$$

where $P\,K1$ *is the* stress acting on the unit of contacts - surface of material particles at the end of the first stage of their pressing, $g/s.m^2$;

$W1$ - pore volume in 1 gram of dispersed material at the end of the first stage of its pressing, cm3;

w_{01} *is the* volume that would be occupied by pores in 1 gram of free bulk dispersed material, to break strong 3 surface contacts, cm

The value was named "aggregation index of dispersed material". Its dimension corresponds to the uniformity of work - g/cm.

2.2 Evaluating the effectiveness of surfactants in the grinding of solid materials

When studying the process of grinding various types of solid materials in a laboratory batch ball mill, it was found by the authors [37] that at the initial stage of the process, the dependence of the change in the content of particles over 80 microns on the grinding time is close to a straight line and, therefore, can be described by the following equation

$$\Delta R008 = K\,t \quad (1)$$

where $\Delta R008$ is the change in the content of particles with a size greater than 80 microns in the material pulverised over time t, %.

K - coarse grinding rate constant, %/min.

As the dispersity of materials increases, the speed of the grinding process decreases, this is due to the strengthening of the surface interaction of the particles to be destroyed, as well as the fact that the reduction of particle size in them reduces the number of defects contributing to deformation (pores, cracks, etc.).

The dependence of $\Delta R008 = f\ (t)$ increasingly deviates from the rectilinear one, and it was found [37] that for all investigated materials the deviation of this dependence from the rectilinear one occurs almost simultaneously with a sharp increase in the surface interaction of the destroyed particles.

Finally, for some dispersibility of the ground materials, increasing the grinding time does not lead to a decrease in the content of particles larger than 80 μm. ie.

$$\Delta R008 = \Delta R008max \quad (2)$$

where $\Delta R008max$ is the maximum change in the content of particles of size more than 80 microns in the ground material, which can be achieved under the given conditions of its grinding process, %.

On the basis of these basic provisions of the kinetics of the process of steam grinding of materials, the following empirical equation can be used to describe it

$$\Delta R008 = \dots$$

(3)

where A_o and A *are* aggregation indices of the material, respectively, coming to the grinding and pulverised during time t, g.cm.

This equation shows that in the initial stage of comminution at small t, when surface interactions of destructible particles are small and comminution is insignificant $(A_o/A=1)$, *the* dependence $\Delta R008 = f$ *(t)* approaches to the form $\Delta R008 = K\,t$

As the material particles change, their surface interactions increase. The value of A_o/A decreases, hence the value of $\Delta R008max \cdot A_o/K \cdot A$ decreases. Finally, after a certain grinding time, the change in the content of particles larger than 80 μm in the material to be ground practically stops. In this case $\Delta R008max \cdot A_o/K \cdot A < t$ and equation (3). takes the form of

$$\Delta R008 = \Delta R008max \qquad (3a)$$

The value of the coarse grinding rate constant K at unchanged mechanical setting of the <u>mill</u> depends only on the individual properties of the particles to be crushed (porosity, microstructure, etc.) and does not depend on the magnitude of the surface interaction. Therefore, the constant K can be used to evaluate the effect of adsorption reduction of the strength of material milled in the presence of surfactants. The more intense the surface interaction between the particles of the material being ground, the smaller the maximum change in the content of particles larger than 80 μm, which can be achieved in the mill regardless of the grinding speed at the coarse stage. This value $\Delta R008max$ can be used to evaluate the influence of surfactants on the stage of the process of grinding of materials, the kinetics of which is predetermined by surface interactions of the particles to be crushed.

When dispersing under the action of external forces, the condensed matter first undergoes volumetric deformation (elastic and plastic deformation) and only then it is destroyed at a certain force [38]. Thus, the work required for dispersion can be divided into two parts. One part of the work is spent on the volumetric deformation of the body and the other part is spent on the formation of new surfaces. The work of elastic and plastic deformation is proportional to volume

$$W_{деф} = K\,V \qquad (4)$$

where K *is the* proportionality coefficient equal to the work of volumetric deformation of a unit volume of a condensed body;

V is the volume of the body.

The work of formation of a new surface during dispersing is proportional to its perturbation:

$$W_n = \sigma \Delta S \quad (5)$$

where σ *is the* surface unit formation energy, or surface tension;

ΔS - the surface increment, or the area of the resulting surface.

The total work expended in dispersing is expressed by the Rebinder equation:

$$W = W_{деф} + W_n = K\, V + \sigma \Delta S \quad (6)$$

Since the volume deformation is proportional to the volume of the body, and $V \sim d^3$ *(d* is the linear dimension of the body) and the surface change is proportional to its initial surface, and $S \sim d^2$, then

$$W = K_1 d^3 + K_2 d^2 \sigma = d^2 (K_1 d + K_2 \sigma) \quad (7)$$

Where K1 and *K2 are the* proportionality coefficients.

From the relation (7) it follows, at large dimensions of the body (at large values of *d) it* is possible to neglect the work of surface formation, then

$$W = K_1 d^3 \quad\quad (8)$$

i.e. the total work of dispersion is determined mainly by the work of elastic and plastic deformation. To calculate the work of crushing, as the first stage of destruction of relatively large pieces of material, we can use this relationship.

At small values of *d,* equation (7) transforms into the relation

$$W = K_2 \sigma d^2 \quad (9)$$

as in this case the work of volumetric deformation can be neglected. The finer the material to be dispersed, the better ratio (9) must be fulfilled. Thus, this relation can be used to determine the work of grinding - the second stage of dispersion At this stage, as follows from relation **(9). the** total work of dispersion is determined mainly by the work of formation of a new surface, i.e. the work of overcoming cohesive forces.

During crushing and grinding, materials are destroyed primarily in places of strength defects (macro- and micro-cracks). Therefore, the strength of particles increases as they are crushed, which is usually Used to create stronger materials, as they are crushed, leads to greater energy consumption for further dispersion.

The fracture of materials can be facilitated by utilising the Rebinder effect, which is the adsorptive lowering of the strength of solids, as discussed above [35].

Surfactant properties. Parameters characterising the efficiency of surfactant application as a grinding intensifier

According to the works of P.A.Rebinder and his school, the basis of lowering the strength of solids in a surface-active medium is the reduction of surfactant molecules surface energy of dispersed materials, as indicated above, also reduces the magnitude of surface interaction between their particles. Therefore, the ability to decrease the surface energy of dispersed materials is an important property of surfactants, predetermining the effectiveness of their use as intensifier of grinding.

In [11] it was shown that the value of their adsorption activity D on the surface of the body can serve as a parameter indirectly characterising the ability of a surfactant molecule to reduce the surface energy of the body:

$$D=\frac{G}{C} \quad (10)$$

where G is the value of surfactant adsorption on the solid surface at formation of a monomolecular layer g (surfactant)/g (solid);

C is the concentration of surfactant in solution at which adsorption monomolecular layer is formed on the solid surface g (surfactant)/g (solid).

This equation shows how many times less surfactant molecules can be adsorbed by monomolecular coating of the solid surface compared to their initial content in the solution before adsorption.

On the surface of real solid solids, there are always micro-pores and cracks with cross-sectional dimensions of surfactant molecules.

When adsorbed, surfactant molecules cannot penetrate into such micro pores and cracks and, therefore, produce an adsorptive strength reduction effect in them. The ability of surfactant molecules to cover the surface of a solid is usually estimated by their area occupied on a given solid when the monomolecular layer is filled $(S_{эф})$ [35]:

$$S_{эф} = \frac{S\,M}{G\,N} \quad (11)$$

Where S is the specific surface area of the solid (usually found from low temperature adsorption of nitrogen),

N *is* Avagadro's number;

M is the molecular weight of the surfactant.

The value Sef is often called the effective area of the surfactant molecule on the surface of a solid, its dimension A

2.3.Features of surfactants and functional composition as grinding intensifiers

According to the nature of adsorption and the mechanism of stabilisation of disperse systems, surfactants are divided into 2 major classes:

niekomolekulirnye and high molecular weight. The first class includes compounds of diphilic character, containing a hydrophilic part and a hydrophobic "tail". The hydrophilic part of the surfactant consists of one or more polar groups: -OH, -COOH, $_{-NH_2}$, -COOMe, -OSO2H;

The hydrophobic one is an aliphatic chain. The second class includes high molecular weight surfactants with alternating hydrophilic and hydrophobic groups. According to chemical structure surfactants are divided into ionogenic and non-ionogenic. Ionogenic surfactants are divided into cationic and anionic. Non-ionogenic substances contain non-ionising end groups with high affinity to the dispersion medium.

The properties of surfactants depend on a number of factors: the mass of surfactant molecules, the arrangement of atoms, bonds, and the forces of interaction between molecules and atoms. Three types of properties are distinguished:

- collective, depending on the total number of molecules;
- additive, the value of which is determined by the sum of the values of properties of individual atoms or groups of atoms in surfactant molecules;
- constitutive, due to the presence of certain atoms
or groups of them and their arrangement in the molecule.

Knowledge of adsorption and surface activity patterns allows a deeper understanding of surfactant properties and their targeted use in building materials technology.

In surfactant adsorption it is necessary to know how the individual functional groups of the molecules interact with the different phases. Additive properties play a major role in this process.

The main properties of surfactants are their ability to lower the surface tension of heterogeneous systems, in particular aqueous solutions. Adsorbed, surfactant replaces the polar surface with a less polar layer, equalising the polarity differences of the phases in accordance with the so-called polarity equation rule.

To intensify the grinding process of cement raw materials for cement production substances: hydrophilising and hydrophobising [37 - 40].

The first group includes sulphite-alcoholic bard (SAB) and its derivatives - sulphite-alcoholic bard (SAB), some esters.

The second group includes soap-soap, asidol, asidol-soap-soap, organosilicon liquids GKZh-94, GKZh-10, GKZh-11, cube residues of synthetic fatty acids, wood pitch and others. Hydrophilising surfactants adsorbing on cement particles keep rather thick layers of water near themselves. This creates a hydrodynamic lubrication between the solid particles, which reduces friction.

The molecules of hydrophobising surfactants with their non-polar part are fixed on the surface of the solid phase, which impairs the wetting of cement with water.

Laboratory experiments and industrial practice have established a significant intensifying effect of surfactants on cement grinding. Thus, Higerovich, Leibovich and Mukhamedzyanov [41] using as a surfactant small doses of soap and oil (0.3 -0.5% of the weight of cement), increased the productivity of cement mills up to 15% with a constant residue on the control sieve. Skromtaev, Rojak and Malinin [40] found that the introduction of sulphite-alcoholic bard (0.3% on dry matter) increased the productivity of cement mills by 4 -15% at constant fineness of grinding, and in the case of constant mill productivity - increased the fineness of grinding from 20 to 50% [43]. Salidzhanov [44], applying as a surfactant soapstock in the amount of 3.075 -0.15%, achieved an increase in the productivity of the mill by 30; As an intensifier of cement grinding used after alcoholic streaming bard - the waste product of molasses digestion on ethyl alcohol [45].

When 0.05 -0.1% of cement weight is added, the productivity of cement mill is increased by 25%. Addition of dioxane production waste and additionally 5 - 30% of adipic acid production waste - fused decarboxylic acids [42] are sprayed with a spray nozzle or by means of a dropper on the material layer in the amount of 0,0,0,25 -0,5%. The productivity of mills is increased by 5 -8%. Additives - wastes of chemical production - scrubber paste of alkyl sulphates, fchotoreagent VZhS and plasticiser adipinovoy PASch -1 [43] allow at productivity of mills at the same level to increase strength of cement on 7 -25%. At the same time the specific by 570 cm^2 /g and by 519 cm^2 g and the residue on the sieve No. 008 will decrease by 1.7 and by 3.1%. Addition of dioxane production waste [44] in the amount of 0.03 -0.1% of the cement weight increases the mill productivity at the grinding of Portland cement by 27%, and at the grinding of Portland slag cement by 12%. Addition of the product of production of higher fatty alcohols of cube residues of isododecyl alcohol production and additional product of production isopropyl phenyl - and -phenylindiamine [45] in the amount of 0,25 - 2,25 of cement weight increases productivity of grinding unit by 40-45%. Addition of soap and water and additionally containing water-soluble polymer of sodium salt of salicylic acid with formaldehyde [46] in the amount of 1 -2,2% of cement mass increases the mill productivity by 25 -30%. The influence of grinding intensifiers based on polycondensation products of amines, as well as known additives (triethanolamine, etc.) was studied [47] It is shown that at the optimum degree of polycondensation the additive based on amines is more effective than the known ones. Introduction of the additive in the amount of 0.2% of clinker mass in the form of a solution of 10% concentration significantly reduces energy consumption and decreases the required duration of grinding. The considered additives practically do not affect the setting time and activity of cement.

The influence of various organic additives (TEA, DEA, ethylene glycol, urea, coal, glycerol, Celex and (T-96) [48] on the grinding process of Portland cement clinkers in ball mills was investigated. Clinker, kaolin and gypsum were loaded into the laboratory mill in the ratio 80:15:5, 20% aqueous solution of additives was added, ground for 75 min. after which the specific surface of cement, energy consumption, etc. were determined. The consumption of additives was 0.01 -1% (per solid).

In a number of experiments grinding was carried out up to a given specific surface area. The duration of the process was determined. It was found that the introduction of most of the studied additives increases the specific surface of cement by 10 -15%, the introduction of urea by 25%. At the same time the energy period decreased by 8 -10 and 20%. Positive influence of additives on grinding process is explained by prevention of clinker agglomeration and reduction of its adhesion to grinding bodies due to neutralisation of free bonding of particles by vapours of organic liquids. It is noted that when the optimum dose of additives is exceeded, their positive effect on clinker grinding decreases. As technically and economically justified consumption of most additives (TEA, T -96 "Celex", etc.) is recommended 200 -300 g per T of clinker.

Grinding intensifiers not only increase grinding fineness and mill productivity, but also influence the quality of the ground cement and its hardening strength [49]. This influence depends on the type of grinding intensifier and its dosage. Usually, the accepted dosage of the grinding intensifier from 500 to 1000 g/t does not adversely affect the strength of cement.

At increased dosage (№ 2000 g/t) cement strength in early hardening period (1 day) can be only 50-70% of its value achieved at the optimum dosage of the grinding intensifier. Alkyl sulphates and alkyl sulphonates have a particularly negative effect on strength. The conditions and duration of storage are of great importance for cement powder and the development of its strength. Storage of cement powder in isolated air conditions for 1 year does not affect the strength significantly, while storage in the open air leads to a significant deterioration of fluidity and reduction of cement strength, primarily in the early hardening period. It is shown that it can somewhat reduce the unfavourable effect of long-term storage of cement under normal conditions on its strength.

Analysis of the literature review indicates the need to search for new cheap and effective intensifiers of grinding on the basis of waste products in order to reduce energy consumption in cement production. In this regard, this work is devoted to the study of the possibility of using wastes from the oil and fat industry of Uzbekistan as grinding intensifiers.

Chapter 3

Cement grinding intensifier

3.1 Preparation of grinding intensifiers.

Gossypol resin (cotton tar), a cube boe residue of cotton oil production, was used for obtaining grinding intensifiers. Cotton sapstock contains neutral fat, fatty acids, gossypol and products of its oxidation, interaction with proteins, phosphotides and fatty acids. The presence of gossypol and its transformation products in cotton soapstock makes it difficult to use it in the soap industry due to unpleasant complications in soap storage (darkening).

To separate fatty acids from the above products, crude fatty acids isolated from cotton soapstocks are distilled.

The cube residue from this distillation is gossypol resin (cotton tar).

The yield of gossypol resin in relation to crude fatty acids extracted from cotton soapstocks is 18-20% [50-55].

Gosipol resin (GS) is a homogeneous viscous mass of dark brown to black colour, insoluble in water, soluble in petroleum distillation products and organic solvents.

Gossypol resin contains from 52 to 64% of free fatty acids and their derivatives. The rest are products of condensation and polymerisation of gossypol, as well as products of its transformation, formed during oil enlargement, mainly obtained in the process of distillation of fatty acids from soapstocks [56].

Up to 15,000 tonnes of gossypol resin per year can be obtained from all soapstocks obtained during the refining of cotton oils, based on the distillation of fatty acids.

Table 1.   shows the parameters of gossypol resin
Kattakurgan Oil and Fats Plant.

Table -3.

Indicators	Kattakurtan MZhK	
	I mode	II mode
% organic matter	97,29	98,66
% inorganic matter	2,71	1,34
% petrol-soluble	-	-
% fat-soluble substances	100	100
% water-soluble substances	-	-
Acid number in mg KOH	65,3	5,0
iodine number (Galus)	99	53
Saponification number in mg KOH	199	171.7
Ether number in mg KOH	134	121,7

Hydrophilic number	91	84,2
Molecular weight (by the cryoscopic method of p -tel'-	525	-
% acetone-soluble part	77.5	83,0
% acetone insoluble part	22,5	17.0
% fatty acids released during saponification	64,0	61,0

The high ether numbers given in this table indicate the presence of lanthanes in gossypol resin [55-58]. The large difference in saponification and acid numbers also indicates the transformation of a modified form of gossypol by the action of alcoholic alkali.

Given the presence of nitrogen, it is possible to assume the presence of condensation products of gossypol with protein substances. The molecular weight of gossypol resins shows that they contain copolymerisation products.

In terms of quality, gossypol resin must fulfil the following requirements.

Table 4.

Appearance	View 1	View 2
	Homogeneous viscous mass from dark - brown to black	
Acid number, mg KOH	70 -100	50 -70
Solubility in acetone, % at least	80	70
Ash content, % at least	1.0	1.2
Moisture and volatile matter content,	4.0	4.0
% max		

Treatment with triethanolamine was carried out to modify gossypol and increase its water solubility. triethanolamine technical

On interaction with ammonia, ethanolamine oxide forms ethanolamines [59]:

$$_2HC - CH_2 + NH_3 = H_2\text{-}N\text{-}CH_2\text{-}CH_2\,OH$$

$$2\,_2HC - CH_2 + NH_3 = H_2\text{-}N \Big\langle \begin{matrix} CH_2\text{-}CH_2\,OH \\ CH_2\text{-}CH_2\,OH \end{matrix}$$

$$2\,_2HC - CH_2 + NH_3 = H_2\text{-}N \Big\langle \begin{matrix} CH_2\text{-}CH_2\,OH \\ CH_2\text{-}CH_2\,OH \\ CH_2\text{-}CH_2\,OH \end{matrix}$$

Triethanolamine

Triethanolamine technical is fire hazardous, low toxic, has alkaline properties [56]. Flash point is 232 0c, ignition temperature is 395°C. Triethanolamine technical on the degree of impact on the human body as a moderately hazardous substance belongs to the 3rd class of danger according to GOST I2.1.007 -76. Triethanolamine technical must meet the following requirements in terms of quality:

Table 5.

Appearance, colour	I C liquid colour yellow to light cinnamon, not darker than aqueous iodine, concentration 1 g/L	II C liquid colour yellow to light cinnamon not darker than aqueous iodine, concentration 15 g/L
Density at 20°C, g/cm3	1,095 -1,124	1,095 -1,135
Fractional composition at residual pressure. 20 mmHg distilled, "% (a) at temperature up to 170°C not more than	14,5	15,0

TEA fulfils these requirements

In the experiments performed to find out the effect of additives on the grinding process of cement clinker, Portland cement clinkers from Ahangaran cement plant and the

Angren Construction Materials Plant of Ahangaran field.   limestone
The results of chemical and mineralogical analyses of cements
are given in Table 4

Chemical and mineralogical composition of cements

Table 6.

Name cement	SiO_2	Al_2O_3	Fe_2O_3	CaO	MgO	BaO	H	n	P
Akhangaran Cement Plant	2,4	4,85	4,27	46,3	1,6	0,16	0,9	2,4	1.17
Angren Construction Materials Plant	2,1	3,87	4,26	45,7	1,4	1,2	0.9	2,45	1,23

Chapter 4

Research methods

When performing experimental works, along with the generally accepted methodology and widely used in practice equipment, some special devices and research methods were used.

All instruments and laboratory equipment used in the research have been checked by the Republican Laboratory of the State Supervision Committee of Standards, Measures and Measuring Instruments.

Determination of grindability of clinkers and limestone was carried out on a laboratory two-chamber ball mill 40T-ML as follows: limestone or clinker was pre-crushed in a jaw crusher and sieved through a sieve to select its fraction size 2 -7 mm. The additive was introduced into the mill before the start of grinding, in the form of an aqueous solution in the amount of 0.015 to 1% of the weight of dry matter (IN TERMS of dry matter) in a finely dispersed (atomised) state.

The grinding duration, the amount of material to be ground (8 kg, 4 kg in each chamber) and the weight of the grinding bodies (32 kg) were always constant.

After the set grinding time, the residual material on the control sieves and the specific surface area were determined.

Plasticising ability of surfactant additives was determined by the change in the melt cone of cement mortar 1:3 on a standard shaking table according to GOST 310-76.

Mixing of cement mortars was carried out in a laboratory stirrer type ML-IA with the number of bowl revolutions equal to 20. Normalisation of cement mortar beams with the size 4x4x16 cm was carried out on the vibrating platform type 435 A with the amplitude of vibrations 3000 -200 per minute.

Bending test of beam specimens was carried out on the device MII-100. Determination of the compressive strength of the specimens was carried out on a hydraulic press PT-100A with the ultimate load in the range of 1.0 - 2.5 tonnes.

The numerical values of the strength of cement mortars given in the tables and figures are arithmetic averages of 4 -6 series of experiments.

In addition to these studies, the surface and bulk properties of aqueous solutions of the obtained surfactants were determined by stalagmometric and specific conductivity methods.

Thus, the research methods used in the work corresponded to the typical modern laboratory analysis and allowed to consider the obtained experimental data as sufficiently objective.

Obtaining new surface-active substances intensifiers of limestone grinding and Portland cement clinkers; Concretes, cement mortars and products made of them are capillary-porous bodies and by their nature are hydrophilic, i.e. being in contact with water, they absorb it. The consequences arising from the harmful effects of water, as well as from the alternate freezing and thawing of moistened concrete in some rare cases become apparent after 5-6 years.

An example is the characteristic failure of kerbstones separating the carriageway from the pavement. But sometimes we see an extreme degree of concrete deterioration due to the action of water.

It is quite natural that defects in concrete from the harmful effects of water do not immediately reach a dangerous limit, but accumulate gradually.

Thus, it is often necessary to prevent the diffusion of water into the concrete body to a greater or lesser extent and thereby reduce the harmful effects of moisture. This is achieved, in particular, by hydrophobising the concrete by introducing hydrophobising surface and aqueous substances.

As a surface-active additive in the production of hydrophobic cement, products from vegetable raw materials and industrial wastes (oleic acid, soap and oil, soapstock, oxidised risikl, etc.) are usually used. Cement with these additives, along with positive properties, is characterised by a number of disadvantages, first of all, increased dust formation during grinding and transportation and excessive air-absorbing capacity in concrete and mortars The purpose of this work is to try to establish the main directions of elimination of these disadvantages by introducing hydrophobic additives of additional components.

Since the gkdrophobising surfactant is adsorbed on dislocations of clinker grains [57-61], its introduction during clinker grinding:

I) increases the probability of grinding particles containing traces of dislocations in the surface layer and leads to the formation of a large number of fine fraction (less than 10 μm) with a high ratio of grain surface to volume, the consequence of which is an increase in sawing; 2) at surfactant dosage exceeding the adsorption capacity of dislocations, part of the surfactant remains on the outer surface of the particles and is a lubricant of the contact surface, which leads to slippage of particles, reducing the time of stay of the material in the mill and reducing the efficiency of grinding. Excess surfactant, which causes lubrication of the contact surface, leads to a "sharp increase in the attraction. 3 In fact, even at the dosage of hydrophobising surfactant corresponding to the adsorption capacity of dislocation, the increase of marginal wetting angle of cement surface leads to the increase of air bubbles adhesion rate to it, and the decrease of surface tension at the boundary solution-air - to the increase of foam volume formed during mixing of concrete and mortar mixtures.

In turn, the mineralisation of air bubbles by the hydrophobic cement particles contributes to an increase in foam resistance. All this strengthens the air attraction when mixing hydrophobic cement with water. Elimination of the mentioned negative phenomena in the production of hydrophobic cement can be achieved by developing polycomponent surface-active additives. In particular, the following measures can be taken to prevent increased dust formation and air entrainment:

1   The surfactant in the state of physical adsorption should be a non-ionogenic surfactant, which has a lubricating effect but is not capable of chemisorption.

2   .Non-ionogenic surfactants will not enhance the blockade of the potential surface layer if they contain only short-chain carbon hydrogen radicals.

3   Weakening of the surface blockade created by the main hydrophobic component of the surfactant can be achieved by introducing a cationic surfactant with polar groups characterised by high dipole moments into the complex additive.

Thus, the polycomponent hydrophobising surface-active additive should contain: the main hydrophobising component necessary for blockade of the potential surface layer in the amount corresponding to the chemisorption capacity of dislocations on the given surface, solvent or ionotenpoic or cationic surfactant.

It is known that surfactants that increase the energetic heterogeneity of the surface and give it a lyophobic-lyophilic character are effective in improving wetting.

The choice of cationic non-ionogenic surfactants for this purpose is based on the high H-bonding energy of the former with water and the highly lyophobic-lyophilic balance of the latter. Wetting agents contribute to "non-extinguishing" and therefore weaken air-involvement. Solvents are necessary for the reason that some hydrophobisers, for example, higher fatty acids (HFAs) with the number of carbon atoms in the main chain more than 10, are viscous substances at room temperature. Their incorporation into cement under industrial conditions is possible by emulsification, heating or dissolution.

Gossypol resin (GS) modified with technical triethanolamine was selected as the main hydrophobising component. Modification allows to change the structures of gossypol resin and molecularity by introducing new polar groups.

Modification is carried out in laboratory conditions in the following weight ratios: HS:TEA - 70:30, 60:40, 50:50 (wt.%), corresponding to the conventional designations I -1, I -2, I -3.

In the reaction flask pour gossypol resin with temperature 50 -60 ° C, then technical triethanolamine, the mixture is heated to a temperature of 100 ° C and kept for 1.5 h with vigorous stirring.

As a result of chemical interaction of components the finished product is obtained, which is a paste-like mass of dark brown colour, well soluble in water.

It is known in the literature that fatty acids R -COOH form salts of the general type with amines P H2. These salts are apparently characterised by the structure of double complex compounds:

$$\left[\begin{array}{c} R \cdot C \overset{.O}{\underset{.O}{}} \end{array}\right] \left[\begin{array}{c} H \quad H \\ N \\ H \quad H \end{array}\right] H \left[\begin{array}{c} R \cdot C \overset{.O}{\underset{.O}{}} \end{array}\right] \left[\begin{array}{c} H \quad H \\ N \\ H \quad R \end{array}\right]$$

Consequently, when fatty acids react with amino alcohols, fatty acid complex salts are formed

R -COOH+ $NH_2C_2H_4OH \rightarrow RCOONH_3C_2H_4OH$

R -COOH+ $NHC_2H_4OH)_2 \rightarrow RCOONH_2(C_2H_4OH)_2$

R -COOH+ $N(C_2H_4OH)_3 \rightarrow RCOONH(C_2H_4OH)_3$

3 our case, the free fatty acids, which are found in gossypol resin, interact with technical triethanolamine to form a fatty acid complex salt of the following type:

R -COOH + $N(C_2H_4OH)_3 \qquad »RCOONH(C_2H_4OH)_3$

The resulting product is water-soluble, has a certain hydrophobic-lyophilic balance and can be used to intensify the grinding process

Study of colloid-chemical properties

A large amount of experimental material has been accumulated, indicating that the change in the properties of disperse systems in the presence of surface-active additives is due to adsorption fixation of additive molecules on the surface of the disperse phase. It depends not only on the molecular state of the additive in solution, determined by lipophilic-hydrophilic balance and hydrophobic interactions of hydrocarbon parts of the surfactant molecule in aqueous medium, but is also a function of the concentration of the additive, its physical-chemical and colloidal-chemical properties.

In this connection we studied surface and bulk properties of aqueous solutions of the obtained surfactants .

Surface energy plays an extremely important role in a large number of very diverse phenomena.

Surface phenomena are expressed in the fact that the state of the molecules in the surface layer is different compared to the molecules in the body volume.

Molecules in the body volume are uniformly surrounded by the same molecules and therefore their force fields are fully compensated. The molecules of the surface layer interact both with molecules of one phase and with molecules of the other phase, as a result of which the reciprocal of molecular forces in the surface layer is not equal to zero and is directed inside the phase with which the interaction is greater. Thus there is a surface tension o, tending to reduce the surface.

Surface tension can also be represented as the energy of transfer of molecules from the body volume to the surface or as the work of formation of a unit of surface.

Surface tension can be expressed as a partial derivative of the Gibbs energy by the size of the interfacial surface at P and $-T$ = const (at constant mole numbers of the components)

$$\sigma = \left(\frac{\partial G}{\partial S}\right) PTnj \quad (12)$$

hence, for an individual substance, the surface tension is the Gibbs energy per unit surface area.

The internal (total) energy of the surface layer $_{and}S$ per unit area is related to the Gibbs-Helm-Goltz equation:

$$us = \sigma - T\left(\frac{\partial \sigma}{\partial T}\right)p \quad (13)$$

where is the latent heat of formation of a unit of surface; T temperature.

Reduction of o can occur as a result of spontaneous concentration in the surface layer of substances with lower surface tension - adsorption.

The value of adsorption is usually expressed in two ways. According to one method, it is defined as the amount of substance in the surface layer A per unit surface area or unit mass of adsorbent (absolute adsorption value):

Or, which is the same thing

$$\Gamma = \frac{v(C0 - Cv)}{S} \quad (14)$$

where C v is the equilibrium concentration of the component in the volume;

C0 - initial concentration of the component in the volume:

V is the volume of the phase.

The adsorption values of the components G and the surface tension are related by the fundamental Gibbs adsorption equation:

$$-dC = \sum_{i}^{E} \Gamma i \, d\mu i \quad (15)$$

$d\mu i$ -chemical potentials of the components.

At low concentrations of C, the relationship goes to the following equation:

$$\Gamma = -\frac{C}{RT} \cdot \frac{d\sigma}{dC} \quad (16)$$

Depending on the sign of the aqueous $d\sigma/dc$, dissolved substances are divided into surface-active ($d\sigma/dc$) and surface-inactive $d\sigma/dc$ and

Surface-active substances (surfactants) are characterised by surface activity, which is a measure of a substance's ability to lower the surface tension at the interface. This value is numerically equal to the derivative ($d\sigma/dc$) , taken with the opposite sign, when the surfactant concentration is reduced to zero:

(17)

Surface surfactant activity is graphically defined as the tangent of the angle of inclination o of the tangent line drawn to the surface tension isotherm at the point of its intersection with the ordinate axis taken with the minus sign.

The purpose of this part of the work was to construct the adsorption isotherm (graph of the dependence $G = f(C)$) of surfactant on the surface of its aqueous solution and to establish the relationship between surface and volumetric properties of aqueous solutions of surfactants obtained on the basis of waste oil and fat industry and pure (technical) triethanolamine.

Surface properties of aqueous surfactant solutions were characterised by surface tension and surface two-dimensional siltation. Volumetric properties were characterised by the value of critical micelle formation concentration (CMC). For this purpose we measured the surface tension of surfactant solutions at concentrations: 0.01, 0.1, I, 2.5, 5.0, 7.5, 10 kg/m^3 . Based on the obtained data, the surface tension isotherm of $\sigma = f(C)$ at constant temperature (293°K) was plotted. The results of the studies are shown in diagram .I.

It can be seen that the surface tension at increasing concentration at first sharply falls, and then this fall slows down and becomes so small that practically the value of o reaches a constant minimum value.

Based on the curve $\sigma = f(C)$, the calculated values of the specific adsorption G, corresponding to different values of the concentration of the substance in the phase from which the adsorption takes place and constructed a curve expressing the dependence $G=f(C)$ diagram 1.

From diagram 1 .it can be seen that the adsorption isotherm of the dependence $G=f(C)$ leaves the origin of coordinates and increases with increasing concentration at first sharply, and then this growth slows down and the curve asymotically approaches some straight line parallel to the abscissa axis, which corresponds to the kind of isotherms characteristic of a typical low molecular weight surfactant.

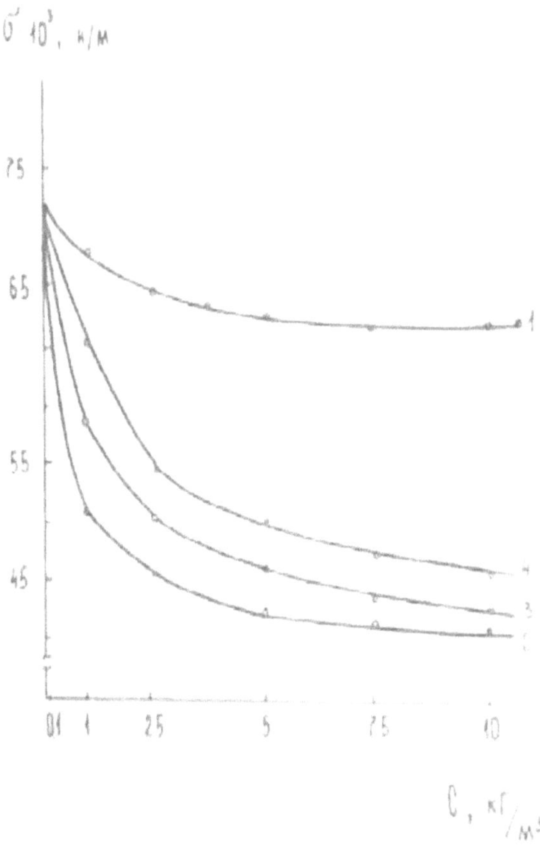

Diagram 1: Isotherms (at 293°K) of surface tension of aqueous surfactant solutions. I) TEA , 2) I -1, 3) I -2, 4) I -3.

50

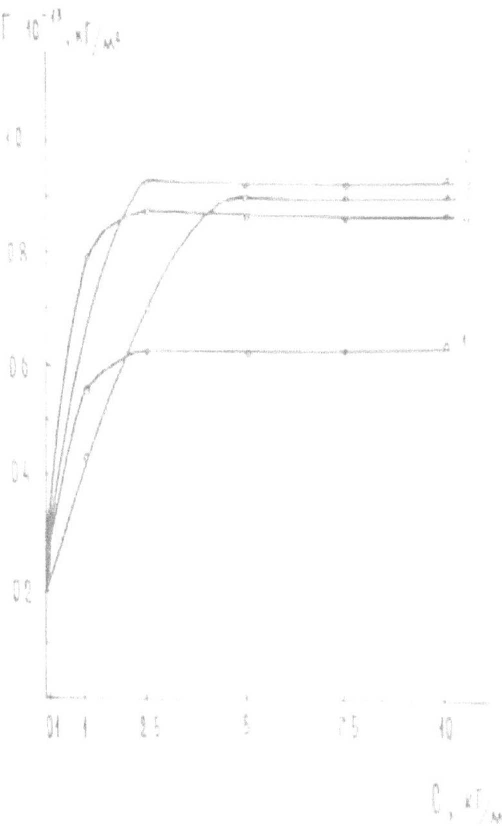

Figure 2: Adsorption isotherms of aqueous surfactant solutions . I) TEA , 2) I -1, 3) I -2, 4) I -3.

The concentration value at which adsorption reaches its highest Gad value and surface tension o its lowest corresponds to complete saturation of the surface layer.

The initial part of the curves at low concentration values corresponds to the surface layer unfilled with adsorbed molecules. As can be seen from the figure, this part of the curves for the surfactants studied by us is almost straight. As the concentration increases, the surface becomes more and more filled with adsorbed molecules and when a certain concentration (10 kg/m^3) is reached, all places on the surface will be occupied by them and the surface tension will cease to change.

Consequently, the adsorption of surfactants is monomolecular in nature, which allows us to calculate the value of Gad, according to the molecular theory of

Langmuir. [Л]

In the Langmuir theory it is assumed that only one layer of molecules can be adsorbed on the interface and therefore the limit value of specific adsorption corresponds to the formation of a saturated monolayer of surfactant molecules on the solid surface.

We calculated the area and length of the adsorbed molecule under the assumption of the usual orientation of the defiled surfactant macromolecules in the adsorption layer. Our calculated parameters were as follows values:

is the cross-sectional area of the polar group:

- the length of the molecule:

$$\delta_0 = \frac{\Gamma_\infty \cdot M}{\rho} = \frac{0,94 \cdot 10^{-13} \cdot 556}{1,09}$$

The appearance of micelles in solution occurs when a certain concentration, called the critical micelle formation concentration (CMC), is reached. The change in the solution structure at the CMC leads to a sharp break in the dependence of its physical and chemical properties on concentration; this fact is the basis of experimental methods for determining the CMC, which is an important characteristic of surfactants.

The critical concentration of micelle formation was determined by the conductivity method.

The resulting data are shown in Figure 2,

The relationship between surface and volume properties was established by equation [87]

$$g = \frac{G_0 \cdot G_{ккм}}{ККМ}$$

The calculations of surface activity value showed that at the ratio of cube residue of cotton oil production of gossypol resin (GS) and pure technical triethanolamine (TEA) (70:30), the surface activity increases much more compared to pure technical triethanolamine (Table 2).

Increasing the concentration of synthesised our surfactants in solution leads to a decrease in their surface activity in accordance with generally accepted patterns for low molecular weight surfactants (diagram 2.).

Two-dimensional pressure of adsorption layers of our synthesised

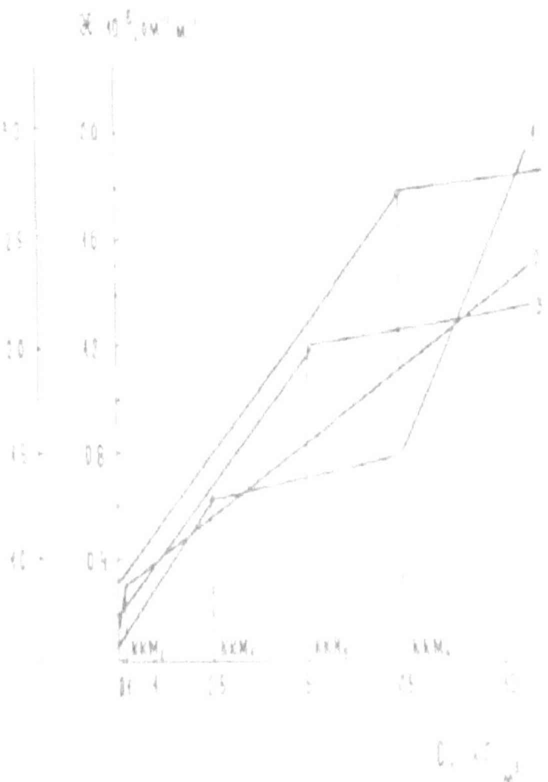

Diagram 3: Isotherm of specific conductivity as a function of concentration of aqueous surfactant solutions. I) TEA , 2) I -1, 3) I -2, 4) I -3.

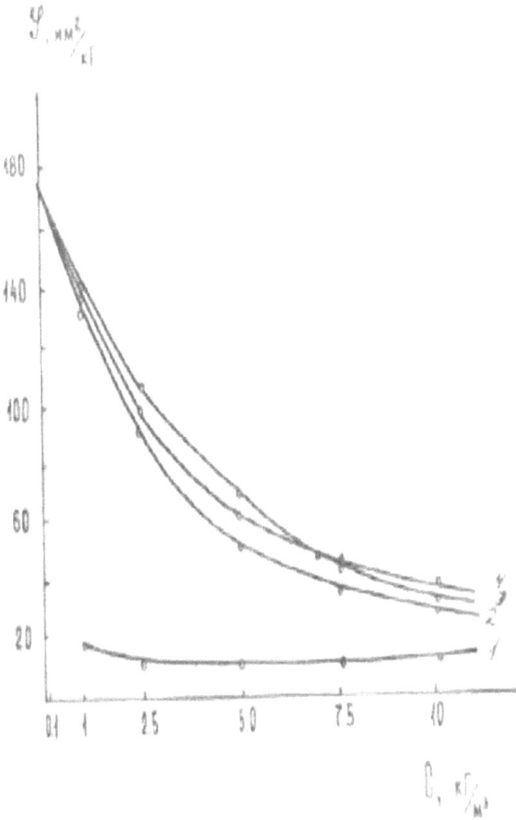

Diagram 4: Isotherms (at 293°K) of surface activity dependence on the concentration of aqueous surfactant solutions. I) TEA , 2) I -1, 3) I -2, 4) I -3.

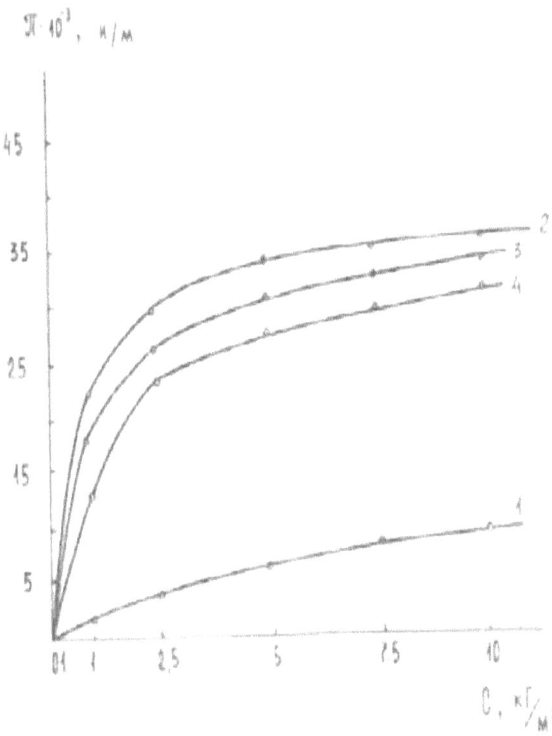

Chart 5.

Isotherms (at 293°K) of the surface (two-dimensional) pressure dependence on the concentration of aqueous surfactant solutions. I) TEA , 2) I-1, 3) I-2, 4) I-3.

Table 5

№	Weighted correlation GS :TEA	Conditional: title	Surface activity, Nm /kg^2
I	70:30	й -1	169 ,00
2	60:40	И -2	60,64
3	50:50	И -3	39,61
4	0:100	TEA	13,24

surfactant is correlated with the value of surface activity . The greater the non-activity, the greater the maximum value of the two-dimensional pressure (Fig.5. curve 2).

Thus, our synthesised substances exhibited surface activity and all the regularities characteristic of low molecular weight surfactants (o, G, T, etc.), which allows us to expect dispersing effect from them

Research methods
4.1 Investigation of the effect of surfactants on the grinding process of limestone and cement clinkers

One of the most urgent tasks of building materials production is to increase the energy efficiency of cement production technology and improve its quality.

The domestic industry is continuously increasing the output of cements, the production of which is associated with grinding to a specific surface of 3700 cm $/g.^2$

It is known that complete grinding of materials is one of the most energy-intensive processes in cement production - it consumes about 80 per cent of all electricity used in cement production. Obviously, therefore, each measure that contributes to the intensification of grinding processes can give a very significant economic effect on the overall scale.

One of the main factors of intensification of the grinding process of solid materials is the use of surface-active substances (surfactants) as intensifiers of grinding.

Effect of surfactants on grinding fineness and specific surface area of limestone

Raw materials of cement production have a variety of physical and chemical properties that influence the process of their grinding. Therefore, in order to improve the performance of grinding and drying plants, it is necessary to study a number of properties of the raw materials used.

For development of technology of dry grinding of limestone with application of surfactants were carried out researches on revealing properties of surfactants, predetermining efficiency of their application as intensifiers of grinding, establishment of parameters, characterising possibility of achievement and size of contribution to intensity of grinding process of adsorption reduction of strength, on studying influence of dosage of surfactants on grindability of limestone.

Grinding was carried out in a laboratory two-chamber ball mill. The grinding time, both without and with additives, was the same - 90 minutes. The additive was fed into the chamber in a finely dispersed (atomised) state. Data on the effect of surfactants on the fineness of grinding and specific surface area of limestone are given in Table 5.1.1.

Influence of surfactants on limestone grinding process performance

Table 7.

Addendum a	Quantity additives, %	Residue on sieve 008, %	Udelnaя surfaces	Specific Expenditure : energies,	Enhancement producers- activities

			strength, p cm $/y^2$	kWh/t	grinding units,%
Without additives		14	2500	41,0	-
И -1	0,03	5,00	2600	33,0	30,0
	0,05	4,75	2650	32,5	32,5
И -2	0,03	5,50	2620	33,0	26,0
	0,05	5,00	2680	33,0	30,0
И -3	0,03	5,50	2600	33,0	26,0
	0,05	4,75	2600	32,5	32,5

The results of sieve analysis show that all the studied surfactants well intensify the grinding process, increase the productivity of the grinding unit. The table shows that the value of specific surface increases with the addition of I -1 at 100 -150 cm^2 '/g, with the addition of I -2 at 120 -180 cm^2 /g, with the addition of I - 3 at 100 cm^2 /g.

When studying the process of grinding limestone with different in the laboratory ball mill, it was found that at one stage of the process, the dependence of the change in the content of particles more than 80 microns on the grinding time is close to a straight line (Diagram 5.), and, therefore, can be described by the following equation [7].

$$\Delta R_{008}=K't$$

As the dispersity increases, the speed of the grinding process decreases. This is due to the strengthening of surface interaction between the particles to be broken, as well as to the fact that as the particle size decreases, the number of defects contributing to deformation (lores, cracks, etc.) decreases. The dependence deviates more and more from the rectilinear one. It was found that for all the studied surfactants the deviation of this dependence from the rectilinear one occurs almost simultaneously with a sharp increase in the surface interaction of the particles being destroyed.

At some dispersibility of the ground materials, increasing the grinding time does not lead to a decrease in the content of particles with a size greater than 80 μm, i.e., [11]. [11].

$$\Delta R_{008}=\Delta R_{008}-max$$

Based on these basic statements of the kinetics of the dry ball grinding process of limestone, the following empirical equation is used to describe it

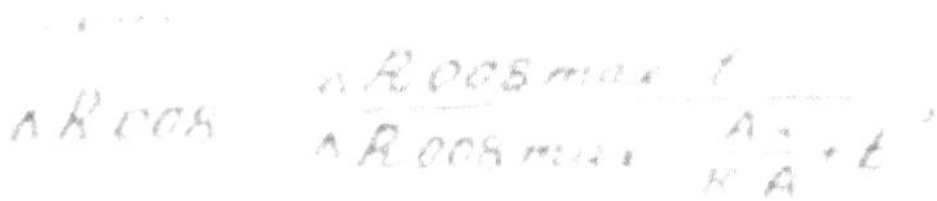

This equation shows that in the initial stage of comminution

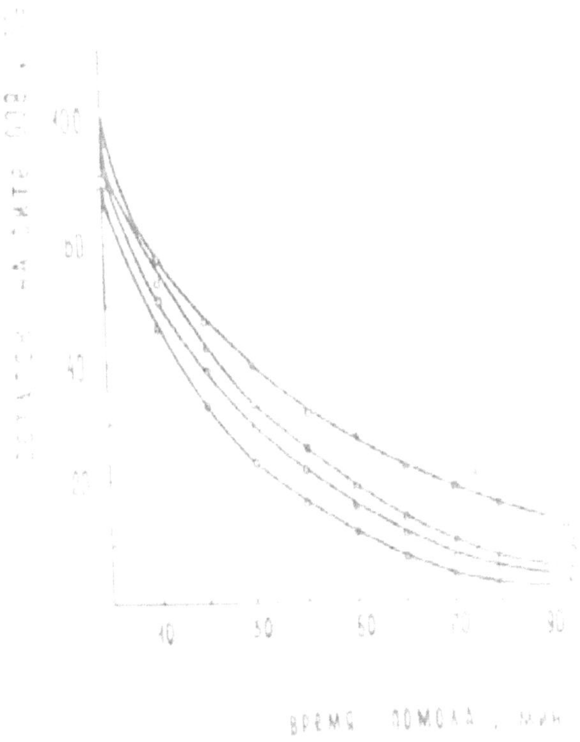

Diagram 6: Curves of limestone grinding kinetics depending on the type and optimal surfactant composition. I) without surfactant, 2) I -1 (0.05%), 3) I -2 (0.05%), 4) I -3 (0.05%).

at small t, when surface interactions of destructible particles are small and change insignificantly ($A_o/A=1$), the dependence $AR0o8{\sim}=f(t)$ approaches to the form $AR0_0\ 8{\sim}=Kt$

As the material particles become finer, their surface interactions increase. The value of A_o/A decreases, hence the value of $AR_{00}\ 8{\sim}maxA_o\ /KA$ decreases. Finally, after a certain grinding time, the change in the content of particles larger than 80 μm in the ground material practically stops. At this $AR_{00}\ 8{\sim}maxA_o\ /KA.{<}t$ and equation (3) takes the form.

$$AR008=AR008{\sim}max$$

59

The value of the coarse grinding speed constant at unchanged mechanical setting of the mill depends only on the individual properties of the particles to be crushed (porosity, microstructure, etc.) and does not depend on the magnitude of their surface interaction, so the constant K can be used to evaluate the effect of adsorption reduction of the strength of the material crushed in the presence of surfactants. The more intensive is the surface interaction between the particles of the material being ground, the smaller is the maximum change in the content of particles larger than 80 μm that can be achieved in the mill, irrespective of the grinding speed at the coarse stage. Therefore, the AR008max value can be used to evaluate the effect of surfactant on the stage of the grinding process whose kinectics is predetermined by the surface interactions of the particles to be crushed. *

Fig.7 shows that surfactant I -1 intensifies grinding more effectively than other surfactants, this shows that I -1 is an effective intensifier of grinding compared to other surfactants.

4.2 Effect of surfactants on Portland cement clinker grinding process

One of the directions of increasing the efficiency of clinker and additives grinding is the development and introduction of physical and chemical methods of intensification based on the creation of adsorption-active medium by small additives of surface-active substances. A common intensifier of grinding in the domestic cement industry is triethanolamine. However, due to limited resources for its production and high cost TEA is not widely used.

The intensifying effect of surfactants forming adsorption-active medium is associated with a decrease in the strength of the ground material at the stage of coarse grinding, reduction of sticking and aggregation at the stage of fine grinding, as well as with an increase in the fluidity (mobility) of cement.

Efficiency of grinding intensifier increases when grinding clinkers with high defect structure and with high adhesion-autohesion properties; it decreases when grinding clinker with easily grindable porous additives, as well as with additives that reduce the degree of adhesion (sand, trepel, slag) with an increase in the moisture content of the charge more than 1.5 -2% [68].

Portland cement clinker grinding was carried out in a laboratory two-chamber ball mill .

During the laboratory experiments, the mill filling factor, the mass of grinding bodies and clinker, its particle size distribution, and grinding duration remained constant. The composition and concentration of surfactants were changed. Grinding time both without and with additives was the same - 90 min.

The additive is fed into the chamber in a finely dispersed (atomised) state. The grindability of clinker without additives and with the addition of triethanolamine, which is widely used in the practice of cement production, is taken as a reference for comparison.

Data on the effect of surfactants on cement dispersibility and on the performance of the grinding unit are given in Table 7.

The effectiveness of surfactant was determined by the quality of residue on sieve #008 and the productivity of the grinding unit based on the correction factor for fineness of grinding [69].

The results of sieve analysis show that all the studied surfactants, as well as the widely used TEA, well intensify the grinding process, increase the productivity of the grinding unit by 30 -47% The optimum concentrations are 0.03 -0.05% of clinker weight.

The analysis of the obtained data showed that the use of surfactants in the above-mentioned quantities gave a reduction in specific energy consumption during grinding to the residue on sieve No. 008 R =10% by 17.5 -25%.

The table shows that the value of specific surface S increases at addition of I -1 from clinker of Ahangaran plant by 350 - -800 cm^2 /g, at addition of I -2 - by 15 -600 cm^2 /g, at addition of I -3 - by 100 -230 cm^2 /g. From clinker of Angren building materials plant the value of S increases at addition of I -1 by 700 - -820 cm^2 /g, at addition of I -2 - by 50 -160 cm^2 /g, at addition of I -3 by 50 -100 cm^2 /g, and at addition of TEA this value is less than that of cement without additives. This is explained by the fact that TEA, adsorbed on the surface of grains, strongly affects the physical properties, increasing the fluidity of cement [70]. Apparently, as the fluidity increases, cement loses its resistance to air permeability. In order to study this phenomenon, we carried out an experiment on grinding cements to a specific surface area of about 5000 with optimal surfactant concentrations (Table 7).

Effect of surfactants on clinker grinding process parameters Table 8.

Name communication additives	Additive concentration, % of weight cement	Dispersibility :cements Z R 008=10%	S cm^2	Specific expenditure e/energy, kWh/t	Enhancement producers- activities of the grinding unit, %
1	2	3	4	5	6
Clinker from Ahangaransk plant without additives	-	10	2650	40,0	-
TEA	0,015	4,25	2600	32,0	37,5
	0,03	4,00	2620	31,5	40,0

1	2	3	4	5	6
	0,05	4,75	2600	32,5	32,5
	0,10	5,50	2640	33,0	26,0
И -1	0,015	4,00	3360	31,5	40,0
	0,03	3,75	3600	31,0	44,0
	0,05	3,50	3650	30,0	44,5
	0,10	4,75	3200	32,5	32,5
И -2	0,015	5,00	2800	33,0	30,0
	0,03	4,50	3200	32,0	35,0
	0,05	4,75	3250	32,5	32,5
	0,10	5,00	3150	33,0	30,0
И -3	0,015	5,50	2750	33,5	26,0
	0,03	4,75	2800	32,5	32,5
	0,05	4,50	2880	32,5	33,0
	0,10	5,50	2700	33,5	26,0
Angren clinker without additives	- -	10	2700	40,0	-

(continued in Table 7.)

1	2	3	4	5	6
TEA	0,015	4,50	2650	32,0	35 ,0
	0,03	4,25	2680	32,0	37,5
	0,05	4,75	2640	32,5	32,5
	0,10	5,00	2600	33,0	30,0
И -1	0,015	4,00	3520	31,0	40,0
	0,03	4,00	3500	31,5	40,0
	0,05	3,75	3600	30,5	44,0
	0,10	4,75	3400	32,5	32,5
И -2	0,015	5,00	2850	33,0	30,0
	0,03	4,75	2860	32,5	32,5
	0,05	4,50	2750	32,0	35,0
	0,10	5,50	2700	33,5	26,0
И -3	0,015	5,50	2700	33,0	26,0
	0,03	4,75	2750	33,0	32,5
	0,05	4,50	2800	32,5	35 ,0
	0,10	5,00	0 278	33,5	30,0

Table 8 shows that the investigated additives intensify grinding well. After 105 minutes of grinding with surfactant additives practically nothing is left on the control sieve, whereas the grinding fineness of cement without additives is 2.3$. The specific surface area of cements with surfactants increases with increasing grinding time.

The additives of the "I" series greatly improve the grindability of clinkers. Clinkers with a diameter of 2-7 mm were ground in a laboratory ball mill. The additives, in optimum quantities from the clinker weight, were introduced immediately before grinding in sprayed form. To determine the residue value on sieve #008, samples were taken from the mill every 10 minutes. The diagrams show the grindability of clinkers without and with additives.

Comparison of the obtained data shows that the effect of the additive on the grindability of clinker, regardless of its mineralogical composition, begins to appear after the first 10 clays, grinding and increases with increasing dispersity of cement.

When grinding cement to the fineness accepted at most plants (8 - 10$ residue on sieve No. 008), the specific power consumption decreased from I -1 by 50 per cent, from I -2 by 45 per cent, from I -3 by 40 per cent and from TEA by 30 per cent.

In addition to these studies, we have carried out work on the effect of surfactant (I -1 on the grindability factor of clinker of Ahangaran cement plant.

There are several ways to intensify the grinding of cement clinker, one of them is the way to intensify the grinding of cement clinker

[71], which consists in the joint grinding of clinker, bi-aqueous gypsum and surface-active additive - polyconsensation product of sulfite modified methacresol-meloaminoformaldehyde - resins in the amount of 0.2 -1% of the weight of cement.

Grinding dispersibility of cements with optimal
surfactant additives
at grinding time.

Table 9.

Additive name	Quantity of additives, %	Variation of dispersibility of cements with optimal surfactant additives with grinding time (min)					
		30	45	60	75	90	105
Akhangaran clinker without additives	-	3350	2750	3500	3900	4500	4800
		18,4	7,5	6,5	5,5	4,75	2,3
TEA	0,03	2000	2650	3550	.3800	4300	4600
		16,7	6,1	6,6	4,9	4,5	0
И -1	0,05	2150	3900	3800	4000	4600	4900
		17 '	7,60	6,4	4,9	3,5	0
И -2	0,05	2150	2850	3850	3950	4600	4800
		17,1	6,8	5,5	4,5	4,5	0
И -3	0,05	3050	2850	3800	3900	4550	4790
		18,1	7,1	6,6	4,8	4,75	0

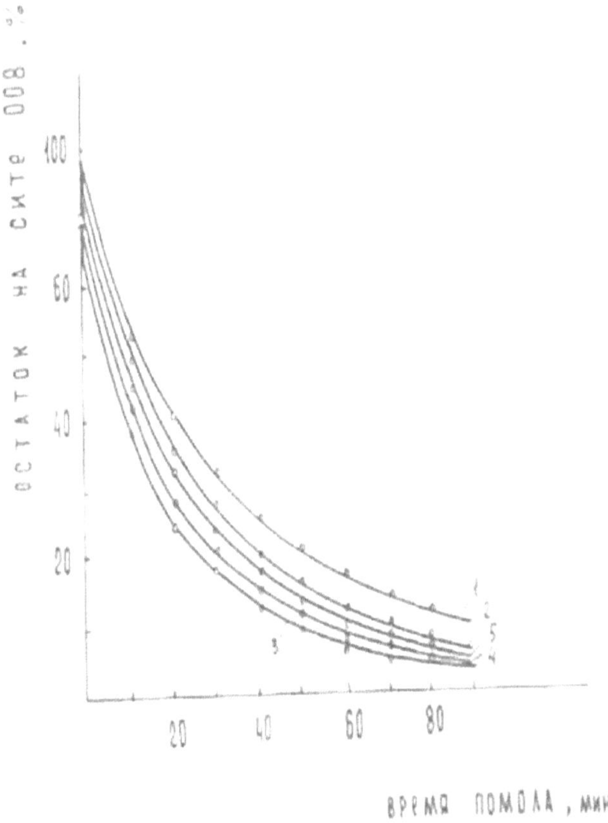

Figure 7: Grinding kinetics curves of portlanddemand clinker from Ahangaran cement plant depending on the type and optimum surfactant composition .
I) no surfactant, 2) TEA (0.03%), 3) I -1 (0.05%), 4) I -2 (0.05%), 5) I -3 (0.05%).

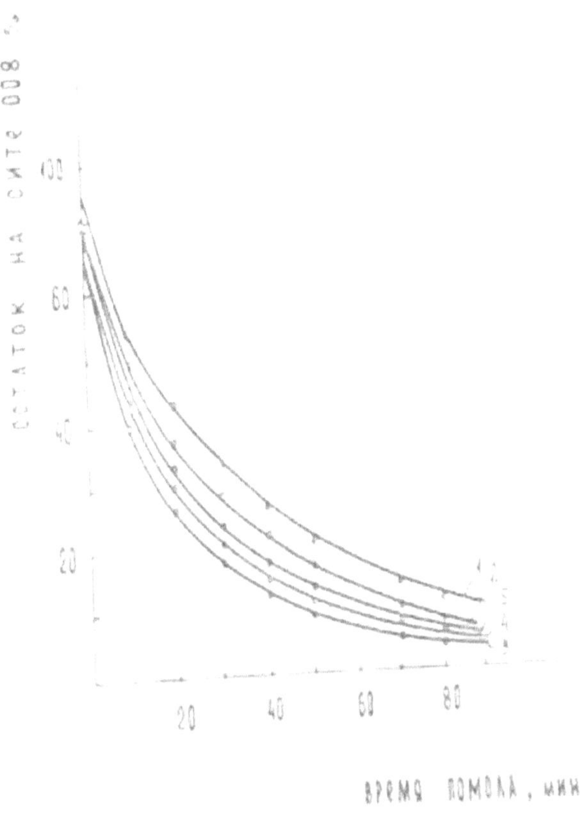

Diagram 8: Grinding kinetics curves of white Portland cement clinker. Angren KCM from the type and optimal surfactant composition. I) without surfactant, 2) TEA (0.03%), 3)I -1 (0.05%), 4) I -2 (0.05%), 5) I -3 (0.05%).

However, the existing method of grinding intensification does not allow increasing the clinker grindability ratio, i.e. the ratio of the grinding time of pure cement to the grinding time of cement with additives at equal specific surface area.

Intensification of cement clinker grinding is carried out in the following way. Triethanolamine-modified cube residue of cotton oil production - gossypol resin (I -1) was introduced together with cement clinker and double-water gypsum into the ball mill in the amount of 0.02-1.00% of the Portland cement clinker weight.

Example I. To obtain Portland cement the following components are mixed, wt.%: clinker 95.0; de-water gypsum 5.0. The mixture is immersed in a

65

laboratory ball mill and subjected to grinding to a specific surface of 3000 cm /g.^{2}

Example 2. To obtain Portland cement mix the following components, wt.%: clinker 94, 085; bi-hydrous gypsum 5,0; cube residue of cotton oil production modified triethanol - amine (I -1) 0,015. Further subjected to grinding similarly to example I.

Example 3. To obtain Portland cement, the following components are mixed, wt.% clinker 94.97; bi-hydrous gypsum 5.0; I -1 0.03. Further subjected to grinding similarly to example I.

Example 4. To obtain Portland cement mix the following components, wt.%: clinker 94,95; bi-hydrous gypsum 5,0; I - 1 0,05. Then subjected to grinding similarly to example I.

Example 5. To obtain Portland cement mix the following components, wt.%: clinker 94,5; bi-hydrous gypsum 5,0; I -1 0,5.

Further subjected to pulverisation in a similar manner to example I.

Example 6. To obtain Portland cement mix the following components, wt.%: clinker 94.0; bi-hydrous gypsum 5.0; I -1 1.0. Then subjected to grinding similarly to example I.

Table 9. shows the grindability of the mixtures according to Examples 1 -6.

Effect of I -1 on clinker grindability factor of Ahangaran cement plant.

Table 10.

By. at measures	Specific surface area, cm /g^2	Grinding time, min.	Grindability factor
1	3000	120	1,0
2	3000	79	1,52
3	3000	77	1,56
4	3050	75	1,60
5	3070	60	2,0
6	3000	57	2,1

The data given in the table show that surface-active additive I -1 allows to sharply increase the cement clinker grindability coefficient. The use of this additive provides the possibility of obtaining in a short period of time cement with a specific surface area of 3000 cm^3 /g, a sharp reduction in specific energy consumption in cement production.

To explain the influence of surfactant composition on cement dispersibility, let us focus on the mechanism of surfactant adsorption on clinker particles. There are two types of active centres on their surface - calcium and oxygen, and the activation of the latter requires the presence of water to ensure their protonation [72]. The active centres previously attributed to silicon and aluminium (iron) [73] are now also considered to be oxygen, the specificity of which is manifested by the presence of incompletely compensated additional valence forces depending on the central metal atom in the oxygen tetrahedron. The main structural defects of clinker particles - crystal boundaries, packing defects, dislocations - are characterised by calcium excess and are mainly zones of calcium centres [74].

Anionactive surfactants, which include all acids and their salts (including sulphonic acids, lignosulphonic acids, fatty acids, etc.), sorbing mainly on calcium centres, contribute to the opening of the main structural defects of clinker particles in cracks, thus anionactive surfactants generate a large number of fine particles during grinding. Increase of fine particle yield (dusting) is characteristic for intensifying action on cement grinding, for example, when introducing such surfactants as common technical lignosulphonates (concentrates of SSB and SB), fatty acids (hydrophobising surfactants), including their water-soluble salts of soap-oil type.

Cationic surfactants, introduced during grinding of binder in the form of aqueous solution, sorbing mainly on water-activated islogenic surface centres, at fine grinding contribute to dispersal of impact or shear energy in defect tips, inhibit their opening into cracks and reduce the amount of fine fraction in cement, increasing the yield of medium fraction (5 -30 μm) useful for strength and other construction-technical properties. Similar results were obtained, for example, in [72-74] for triethanolamine, the most typical of cationic intensifiers of grinding.

As for our intensifiers, these additives, unlike the known ones, are chemisorbed on active surface centres of both calcium and oxygen types. and oxygen [12]. Therefore, we can consider the process of clinker grinding in the presence of I -1, I -2, and I -3 as a result of activation (i.e., opening into cracks) of structural defects enhanced or generated by adsorbed surfactant molecules.

Such structural defects can be divided into two groups - starting on calcium and oxygen atoms, and at the beginning of grinding the number of the former is much higher, thus practically we deal with two different materials - containing initial defects and without them, which leads, as it is known, to the oscillatory mode of grinding. Let us call the grinding mode resonant, when under the influence of surfactant both groups of defects develop into cracks, mutually accelerating the process of destruction of initial grains (looking for early resonance). Obviously, such a mode can be established by regulating the functional composition of the additive, namely, the molar ratio of anion- and cation-active groups in it, the inverse in its value of the ratio of the number of calcium to oxygen centres, which can be calculated from the mineralogical composition of clinker and the density of dislocations on its surface.

Thus, the introduction of additive series "I", intensifying the grinding of the binder, prevents overgrinding and increases the yield of the average fraction in the cement.

List of references used

1. Potapova E.N., Volosatova M.A. Cement Production Moscow 2014.

2. Panova, A. V. Cement production technologies and negative environmental impact / A. V. Panova. V. Panova   :

direct // Young Scientist. - 2022. - № 1 (396). - C. 21 -23. - URL: httpshttps://moluch.ru/archive/396/87638/moluch.ru/archive/396/87638/ (25.10.2023).

3. https://www.avtobeton.ru/materiali dla proizvodstva_cementa.html

4. https://energosteel.com/syre -pri -proizvodstve -cementa

5. Dmitriev A.M., Kaushansky V.V. Problems of utilisation of technogenic materials in cement production. - Cement. - 1988. № 9. - C.2 -3.

6. Kushchidi V.I. Fuller utilisation of wastes and by-products in the industry. - Cement. - 1982. - № 2. - C. 1 -4.

7. Pyachev V.A. Problems of industrial waste utilisation in cement production // Ecological technology. Processing of industrial wastes into building materials; - Sverdlovsk: ed -vo UPI, 1984. - C.4 -7.

8. Materials of the XXU1st Congress of the Communist Party of the Soviet Union. - Moscow: Politizdat, 1985. - 352 c.

9. Khodakov G.S. Physics of grinding. -M.: Nauka, 1972. - 307 c.

10. Deshko B.I., Krainer M.B., Krykhtin G.S. Grinding of materials in cement industry. - Moscow: Stroyizdat, 1966. - 271 c.

11. Nudel M.E. Physico-chemical researches of process of grinding of raw materials at dry method of cement production. Avtoref. dis kand.tehn.nauk. - M., 1977. - 21 c.

12. Timashev V.V., Krykhtin G.S.. Nudel M.E. Intensification of work of grinding-drying plants by introduction of surfactants. - Cement. - 1975. - 1 6. - C.4 -5.

13. Zhmodinova M.S., Sudanas L.G. Mobility of raw material mixtures used in dry method of production. - Cement. - 1974.№ 5. - C.16 -17.

14. Sudanas L.G., Zhmodinova L.S., Leveiches L.A. Mobility of powdery materials. - Cement. - 1977. - L I. - C. 19 -20.

15. Goldstein L.Ya" Intensification of technological process and utilisation of industrial waste in cement production // ed.va.Giprocement 1970

16. Tarnarutsky T.M., Yudovich Y.E., Vatutina L.S. Application of LSTM-2 additive for production of high-strength cements. - Cement. 1984. - № 8. - C. 13 -15.

17. Pirotsky V.Z., Matseev N.S., Demin A.V., Korotayeva Z.M. Technology of cement grinding with the use of intensifiers. - Cement. - 1988. - № I. - C.17 -19.

18. Rojak S.M., Pirotsky V.Z. Resistance of different clinkers and conditions of the grinding process // Tr. Research Institute of Cement. - 1960. - Issue 14. - C.3 -41.

19. Bogne R. // The chemictry of Portland Cement. -1955 IV.P.29 -33

20. Guman V.S. Intensification of grinding of silicate materials with the help of organosilicon compounds. Avtoref.dis. . kand. tehn.nauk. - Kiev, 1969. - 23 c.

21. Khodako G.S. Fine grinding of construction materials. - M. : Stroyizdat, 1972. - 239 c.

22. Tovarov V.V., Gorlinova-Astapovitskaya S.N. Wet grinding of cement clinkers with application of PRS. - Moscow: Proshtroyizdat, 1951. - 91 c.

23. Bacvord A.// Roc produkte. 19.V.12

24. Kinderrett S.. // Chemical ensiklopedia. 1952 -13 -P -35

25. Wagener R.// Zement -Kalk -Gios -1961. -V -10S.5

26. Khigerovich M.I. Hydrophobic cement. - M.: Promstroizdat, 1957. - 208 c.

27. Deryagin B.V., Krotova K.A. Adhesion. - M. -L.: M. 1949 C. 242^

28. Shmidt.A. //Zement -kalk -Gips. - - 1949. -v -17 -P -3.

29. Rlorsot B. //Bottk produkte. 1952. -V12.P.3.

30. Beke B. Oparky J. -Jn: Sumpos. Zerkleinern. Amsterdam. -1966. P.407..

31. John R. Lowe. Microstructural picture of fracture // Proceedings of the International Conference on Solid State Fracture Problems. -M.: Metallurgizdat, 1967. - C.36 -41.

32. Westwood A.R. Influence of medium on the fracture process // Proceedings of the International Conference on Solid State Fracture Problems. -M.: Metallurgizdat, 1967. - C.61 -66.

33. Johnston T.L., Parker E.R. Fracture of nonmetallic crystals // Proceedings of the International Conference on Solid State Fracture Problems. -M.: Metallurgizdat, 1967. - C .96 - -100.

34. Rebinder P.A. VI Congress of Physicists. - Moscow: Gosizdat, 1928.

35. Rebinder P.A., Schreiner L.A., Zhigach K.F. Hardness and Drilling Reducers //Surface Phenomena in Dispersed Systems. Physico-chemical mechanics. - M., 1979. - C.270 -320.

36. Deryagin B.V. Properties of thin fatty layers and their role in disperse systems. - M.: VSNITO, 1937. - 121 c.

37. Nudel M.E., Kryhtin T.S. Features of the process of dry grinding of cement raw materials in a surface-active medium // Proc. of the Cement Research Institute. Cement Research Institute. - 1976. - Issue 36. - C.34 -52.

38. Frolov Y.G. Course of colloid chemistry. Surface phenomena and disperse systems. -M.: Khimiya, 1989. - C. 115 -118.

39. Gregg S., Sing K, Adsorption, specific surface area, porosity. - M.: Mir, 1970. - 407 c.

40. Karibaev K.K. Surface-active substances in the production of binding materials. - Alma-Ata: Nauka, 1980. - 336 c.

41. Higerovich M.I., Leibovich H.M., Mukhamedzyanov M.M. Intensification of cement grinding by small doses of soap and oil // Information messages of the Research Institute of Cement. - 1951. - Issue 20. - C.21 - 25.

42. Skromtaev B.T., Rojak S.M., Malinin Y.S. Production and application of plasticised cement. -M.: Promstroizdat, 1953. - 245 c.

43. Cement concrete with plasticising additives / S.V.Shestoperov, F.M.Ivanov, A.I.Zatsepin, T.Yu.Lyubimova. - Moscow: Promstroizdat, 1952. - 207 c.

44. Salidjanov S.B. Application of soapstock as an intensifier of raw material and clinker consumption and cement hydrophobisation at cement plants. - Tashkent: 1955. - 4 c.

45. A.S. NO. 443008. MKI^3 C04 IN 7/54. Cement grinding intensifier / I.F.Kascheev, Y.B.Porisus, N.M.Endrikson, S.A.Polishchuk . - I p.

46. A.S.. № 1046216.. YUT^3 C04 IN 7/54. Cement grinding intensifier / S.L.Protutan, V.P.Pedan, G.R.Goltsman, G.D.Dibrov, S.K.Kovalev, L.P.Proekutin, V.B.Pokhel . - 3 p: Il.

47. Bryzhik T.G., Eryzhik A.V., Lutinina I.G., Vezhlivtsev V.A. Use of new intensifiers of grinding at Starooskolskiy ZaEOD. - Cement. - 1981. - & II. - C. 17.

48. A.S. NO. 874694. MKI^3 C04 IN 7/54. Method of cement clinker grinding / G.N.Oskalenko, Yu.V.Svetkin, V.A.Kulik, N.Y.Kuzmenko (USSR). - 2 p: Il.

49. A.S. NO. 963967. MKI^3 C04 IN 7/54. Cement grinding intensifier / I.A.Oshchenkov, E.N.Elbert, A.M.Shidanova, V.P.Goncharov, V.M.Dmitriev, I.H.Sharipov, L.V.Barkov, A.Zenkov. - 3 p: Il.

50. A.S. NO. 697433. USSR. MSH^3 C04 IN 7/54. Additive for cement grinding intensification / I.B.Zelenev, L.N.Popov . - 2 p: Il.

51. Kenigsberg Z.I., Brepman A.M. Distillation of fatty acids from - upstocks of black cotton oils // Proc. of the soap sector. Steklographic edition of VNIIZh. - 1951. - C.118.

52. Satgansky M.P., Shvetsov A.S. Distillation of fatty acids from soapstock. - Masloshrovaya promyshlennosti. - 1953. - № 8. -C.24 -25.

53. Zamyshlyaeva AM, Slorina GZ, Belopolsky AM, Bunina V. On the composition of gossypol resins obtained by distillation of fatty acids from cotton soapstock // Tr. VNII fats. of - Vp.24. - P.282 -295.

54. Grsch S. Analysis of fats and waxes. - M. -L.: Gos.chemico-technical izdvo, 1932. - C.208.

55. OST 18 -114 -73. Gossypol resin. - M.: MP SSR, 1973. - 7 c.

56. General chemical technology of organic substances / Edited by D.D.Zykov. - M.: Khimiya, 1966. - C.247.

57. TOU 6 -02 -916 -79. Triethanolamine technical. - M.: MKHP SSSR, 1970. - 6 c.

58. Tarnautsky G.M. Development of technology and research of construction-technical properties of hydrophobic Portland cement with polycomponent additives. Avtoref. dis. kand.khim.nauk. M., 1974. - 23 c.

59. Yakhnin E.D., Taubman L.Y. Physico-chemical mechanics of soils, soils, clays and building materials. - Tashkent: FAS, 1975. - 421 c.

60. Becker B. Introduction to the electronic theory of organic reactions. Moscow: Mir, 1965. - 381 c.

61. Nevolin F.V. Chemistry and technology of CMC. - Moscow: Gostoptekhizdat, - 545 p.

62. Schwartz A., Perry J., Eertsch J. Surfactants and detergents. - M., 1960. - Vol. 2. - 555 p.: Il.

63. Markman A.P. Chemistry of lipids. - Tashkent: Uzbek SSR Academy of Sciences, 1963. VOL.1. - P.31.

64. Tursunov A.K., Fathullaev Z.F., Jalilov A.T., Sadykov *A*.A., Isayeva T.H. Study of fractionation and use of gossypol resin // ZhPH. - 1985. - Т.ХУШ, № 4. - C.878 -880

65. Laboratory works and tasks on colloid chemistry / Edited by Yu.G. Frolov. - M.: Khimiya, 1986. - C .8 -10.

66. Shpykova L.G., Chikh V.I. Genesis of microstructure and properties of cement stone // Hydration and hardening of binders. - Ufa, 1978. - 0,299 -306.

67. Polak A.F., Babkov V.V. Mathematical model of poly-disperse structure // Hydration and hardening of binders. - Ufa, 1978. -C.3 -11.

68. Shestoperov S.V., Ivanov F.M., Lyubimova T.Yu. The action of plasticisers on cements of different mineralogical composition. - Cement, - 1952. -P 6. - P.13 -15.

69. Ratinov V.B., Ivanov F.M. Chemistry in construction. 2nd edition, revision and supplement. -M.: Stroyizdat, 1977. -220 c.

70. Tretyakova A.S., Uryvaeva G.D., Lyvinenko A.T. Influence of some hydrocarbons on the hardening of clinker minerals // Hydration and hardening of binders. - Ufa, 1978. - 0.130.

71. Young J.Y. Influence of Sugars on hydration of tricalcium aluminate //IV International Congress on Cement Chemistry. -M.: 1978. -C.209 -210.

72. Loher F.B., Richard W. Investigation of the mechanism of cement hydration // UP International Congress on Cement Chemistry. - M.: 1976. - Vol.2, Book 2.

- С.122 -133.

73. Rojak S.M. Tamponage cements //VI International Congress on Cement Chemistry. - M., 1976. - T.3. - C.231 -242.

Printed by Books on Demand GmbH, Norderstedt / Germany